THE Hawks Nest TUNNEL
AN UNABRIDGED HISTORY

Dam and tunnel intake site, with nine foot butterfly valves and casings for sluices shown in foreground. *Courtesy West Virginia State Archives.*

THE Hawks Nest TUNNEL
AN UNABRIDGED HISTORY

PATRICIA SPANGLER

PUBLISHING

© 2008 by Patricia Spangler.
All rights reserved.

No part of this book may be reproduced or transmitted in any form, by any means (electronic, photocopying, recording, or otherwise) without the prior written permission of the publisher.

Library of Congress Catalog Card Number: 2007939700
ISBN-13: 978-0-9801862-0-8

First Edition

Printed in the United States of America

Front and back cover images courtesy West Virginia State Archives.

Interviews from *Goldenseal* magazine reprinted by permission.

Selections from *U.S. 1, BOOK OF THE DEAD*:
Reprinted by permission of International Creative Management, Inc.
Copyright ©1938 by Muriel Rukeyser.
From *The Collected Poems of Muriel Rukeyser*, ©2005 Pittsburgh University Press.

Exhaustive attempts were made to determine the copyright holder for *Man on a Road* by Albert Maltz, to no avail. The publisher appreciates any information towards that fact.

All newspaper articles courtesy their respective copyright holders.

Wythe-North Publishing
P.O. Box 1208
Proctorville, OH 45669-1208

www.wythe-north.com

In memory of my mother
Jane Palmer Stephens
1919-2007

TABLE OF CONTENTS

Acknowledgments ix

Foreword x

Introduction xiii

Chapter One: The Congressional Investigation of Hawks Nest Tunnel, 1936 1
                 Statements by: Philippa Allen 2
                                 Emma Jones 27
                                 Charles Jones 31
                                 Hiram Skaggs 39
                                 Arthur Peyton 42
                                 George Robison 50
                                 William Finke 58
                                 Rush Dew Holt 59
                                 John W. Finche 61
                                 James Mason 63
                 Letter of Rebuttal from Rinehart & Dennis Co. 69
                 Subcommittee Findings and Conclusions 73
                 Letter of Opinion, Jennings Randolph 75

Chapter Two: Interviews with Tunnel Survivors 77
                                 Stanley Cavendish 77
                                 B. H. Metheney 81
                                 Louise Harless 86

Chapter Three: Literature Spawned by the Incident at Hawks Nest 91
                                 U. S. 1, Book of the Dead, Muriel Rukeyser 91
                                 Man on a Road, Albert Maltz 104

Chapter Four: Through the Eyes of the Media, 1928-1936 111

Afterword: A New and Greater Regard for Humanity? 181

Appendices:	A. Biographical Information	185
	B. Related Documents	192
	C. Glossary of Terms	249
	D. Glossary of Names	250
	E. Dimensions & Specifications	252
Index		253
Bibliography		260
About the Author		264

ACKNOWLEDGMENTS

From conception to birth, the gestation period for *The Hawks Nest Tunnel: An Unabridged History* has taken approximately ten years. Probably not an inordinately long period for a history of this type, but problematic in that now I must rely upon memory to thank all those who assisted and encouraged me in the process.

With apologies to anyone I've left out, I wish to extend my heartfelt thanks to my sister Jennifer for patiently listening to every challenge and dilemma encountered during this process; to Bill Clements of the West Virginia Book Company for generously providing the photos of the tunnel's construction; to all of the librarians at the Fayette County Public Library in Oak Hill and Fayetteville; to Debra Basham, West Virginia Cultural Center; to Martin Cherniack for his inspiration and support; to the Muriel Rukeyser Foundation for their generosity; to my cousin Anne Harless for providing invaluable editorial assistance and moral support; to everyone else who has expressed interest in this project; and finally my appreciation, gratitude and love to my husband Keith, for patiently tolerating the many, many hours I've spent consumed in the excavation of details from the past.

FOREWORD

Reflecting back upon the process of writing *The Hawks Nest Tunnel: An Unabridged History*, I realize the journey began on the morning of May 15, 1997 as my husband and I stood in stunned silence on top of Cotton Hill Mountain near the rim of the New River Gorge. It was the first day of our long journey to reclaim an abandoned mountain farm. Voluntarily homesteading *a la* Willa Cather *sans* house, running water, electricity or telephone—with our worldly possessions piled at our feet in an unimpressive heap of garbage bags and cardboard boxes.

Leftover wet from the previous night's thunderstorm promised a typical West Virginia day replete with sauna-like humidity and ever-present gnats. I was struggling with the question of "What in God's name have we done?" when the stillness of the morning gave way to a baritone wail mushrooming forth from the belly of the New River Gorge, blanketing the earth with its drone.

Did I ask what it was? I don't remember. But I do remember the brief explanation offered by my husband: "Hawks Nest," and implicit within that abbreviated explanation the message that within minutes a huge volume of water would burst forth from the Hawks Nest dam; released to course once more over through and around the riverbed of the ancient New River. The siren's message warned all within hearing range to immediately seek higher ground; yet sitting as we were, high above the reach of the river's torrent, the dam's release posed no threat.

For a brief moment I indulged in reverie, reliving family picnics shared at the Hawks Nest State Park, replete with the aroma of hot dogs carefully roasted over stone cook-pits built by young Citizen Conservation Corps youth in the '30s—their stonework still lovely today. Hawks Nest, the spectacular overlook where many years ago, during my Wisconsin grandparents' first and only visit to West Virginia, we spent an afternoon overlooking the dam. As I shoved quarters into a gizmo that looked more like an ancient deep-sea diver's outfit than a telescope, John and Albina Palmer, parents to my mother, stood breathless in the face of such magnificent natural beauty.

Sweet memories blipped across the radar screen of my consciousness then vanished as quickly as they appeared, for when *home* refers to a chigger-infested briar patch, survival instincts supersede daydreaming. Thinking back on that *siren moment* I realize it was simply one thread woven into the fabric of that summer, albeit one that became a focal point in my life for several years.

During that first summer and fall, our days were occupied harvesting trees to dry in preparation for the lumber our home would require; setting fence posts and creating pastureland for llamas and goats; and cultivating flowers in an effort to infuse a semblance of *civilization* into this raw, extraordinary experience. Yet through it all, when least expected, the siren would often cry out from the valley below, gradually encouraging me to delve beneath the surface of its plaintive wail until I was forced to admit that, aside from the pleasantries I've mentioned, the depth of my actual knowledge of Hawks Nest, its dam and tunnel, could be summarized quickly: water from the New River flows through the tunnel to exit at the hydro plant on Gauley Mountain; and during the tunnel's construction, countless men were exposed to silica dust and died as a result.

Truly my ignorance was impressive and, perhaps because nature abhors a vacuum, information gradually filtered into this void. Unsought-out articles related to the tunnel's history appeared. During casual conversations friends and relatives offered up tantalizing tunnel lore. With each bit of information my curiosity increased until I was locked in the throes of a focused, passionate quest to uncover everything I could find related to the tunnel's history.

Did I intend to write a book? Not initially, although in retrospect a journal entry from October 7, 1999 reveals how completely my perceptions were colored by the tunnel's history:

> *October 7, 1999: Stripped bare overnight, the earth. Shorn of her garments, her leaves-of-many-colors. Now, simply pure form. Spine, vertebrae, bones threatening to pierce her fragile skin. Ridges, angular and sharp. Ancient streambeds etching space between anorexic ribs.*
>
> *I turn away, shocked at this vulnerability. Thinking of my own mother, I yearn to share this image with her. She would understand my reaction to such essential rawness.*
>
> *I shiver. From the cold, damp wind?—or the memory of last evening's read, too graphic to forget?*
>
> *"…And typically the workers presented a falling off of weight, a considerable amount, in a very short time. The men got down so they had no flesh left on them at all. As they express it down there, the men got so they were all hide, bone and leaders, which means he is just skin and tendons and looks like a living skeleton."*[1]

There is no disputing the fact that the construction of the Hawks Nest Tunnel was truly *a marvel of engineering expertise*, and that aspect of its history is undeniably interesting. Yet the tunnel's hook for me has little to do with the mechanics of its construction and everything to do with the depth and breadth of the human suffering that resulted from its construction.

In pursuit of the tunnel's lore, I spent month upon month gathering documents

---

1. "An investigation relating to health conditions of workers employed in the constructions and maintenance of public utilities: hearings before a subcommittee of the Committee of Labor, House of Representatives, Seventy-fourth Congress, Second session, on H.J. Res. 449.: January 16, 17, 20, 21, 22, 27, 28, 29, and February 4, 1936." p. 28.

and data, scrutinizing microfilm, and studying the political and social environment of the 30s in hopes of understanding both the milieu as well as the motivation and mindset that enabled the tragedy to occur.

At one point, overwhelmed and deeply saddened by my conclusions, I packed up the research and set it aside. But time and again my efforts to ignore the ever-present voice of the siren failed and the unfinished project continued to scratch—sometimes claw—its way into my thoughts. Finally thanks to a friend who shamed me into finishing this project, I revisited it only to realize immediately that the brief hiatus from research had helped facilitate a shift in my perspective.

Initially I was struck by the variety of materials collected during the initial phase of my tunnel research. Clearly my own *tunnel quest* would have benefitted from access to this compilation of material. Likewise it was apparent that this volume of material might also assist others. Finally I realized that my initial motivation—a desire to cast blame, to judge, to vilify—no longer drove this project. Consequently I have consciously refrained from *interpreting* these events; attempting instead to present this history as objectively as possible, with the conviction that the history of Hawks Nest—a smorgasbord replete with ethical, philosophical, social, and cultural offerings of the 30s—is an epic saga quite capable of telling itself.

From the Cotton Hill Bridge to Chimney Corner, WV Route 16 parallels the New River. Today this section of water—referred to in the summer as the New River "dries"—is swollen from recent rains, a kayaker's delight. Red and white signs plaster the trees along this stretch of highway and from a distance look more like misplaced Tibetan prayer flags than warning signs for boaters and fishermen. *Danger, Dam Upstream* the signs announce. Closer examination explains that the siren's blast warns fishermen, boaters, and swimmers alike to *head for higher ground*. Preemptive words of caution made all the more ironic considering the utter absence of health advisories extended to the men who perished during the tunnel's construction.

This spring marks the beginning of our tenth year on Cotton Hill Mountain. Dozens of seasons and a hundred full moons. Seasons of change and growth. Seasons of sadness and loss. No two ever the same. Yet the siren remains ever constant, its melancholy reminder but one thread in the tapestry of sounds that embrace us. Intimate as the wind; seductive as the train's haunting valley-floor whistle; and often as disturbing as the roar of Laurel Creek, when swollen and uncomfortable from heavy rains.

I smile now, remembering. For in fact what I knew about Hawks Nest then was little more than a quizzical "Isn't that where some men died?" What I knew then may have filled a page. What I didn't know fills this book.

INTRODUCTION

During the early part of the twentieth century, while most of the nation teetered on the precipice of the Great Depression, officials of the Union Carbide and Carbon Corporation, New York City, were finalizing plans for an industrial project in West Virginia that would, within a few short years, become known as "America's Worst Industrial Disaster."[2] That industrial undertaking—and the subject of this book—is collectively referred to as the Hawks Nest construction project.

Aided by the synchronistic conversion of conditions unique to the 1920s, financially solvent corporations such as Union Carbide were quick to parlay a ruptured economy, a starving workforce and laissez-faire governmental oversight into fodder for unprecedented corporate profit.

After acquiring huge amounts of land adjacent to the New and Kanawha Rivers in Fayette County,[3] the Union Carbide and Carbon Corporation created a subsidiary, The New-Kanawha Power Company, and proceeded to file papers of incorporation with the State of West Virginia classifying itself as a *public service utility* company.[4]

In reality the New-Kanawha Power Company was simply a legally-construed vehicle created to facilitate acquisition of a permit from the WV Public Service Commission for the purpose of constructing a dam across the New River near Hawks Nest, West Virginia. Although initial relations with the PSC were congenial, soon after the permit was obtained, relations with the PSC turned "south,"[5] and in 1933 Carbide transferred the New-Kanawha Power Company to another subsidiary—the Electro-Metallurgical Company.[6] Finally, in 1935, having long out-lived its initial function, the New-Kanawha Power Company was dissolved.[7] Not one single day of its short existence was spent operating as a public service utility.

The overall scope of Carbide's Hawks Nest project was threefold and included the construction of a dam across the New River at Hawks Nest; the erection of a hydroelectric power plant at the base of Gauley Mountain; and most significantly, the excavation of a tunnel connecting the dam at Hawks Nest with the hydro plant at Gauley Bridge. By way of the tunnel, the flow of the ancient New River would

---

2. Cherniack, Martin. *The Hawk's Nest Incident: America's Worst Industrial Disaster.* Yale University Press. New Haven, CT. 1986.
3. See "Electro-Metallurgical Company Hawks Nest Hydro Electric Development." Appendix B.
4. See "Notice to form Corporation," Appendix B.
5. See "Miscellaneous Correspondance," Appendix B.
6. See "The Public Service Commission of West Virginia.....," Appendix B.
7. See "State of West Virginia," Appendix B.

be diverted from the base of the dam and re-routed through Gauley Mountain to emerge again at the power plant near Gauley Junction. There the river's water would be channeled through four penstocks connected to generators and thereby generate electricity required to fuel the operations of the large Union Carbide production plant at Boncar,[8] five miles further downstream.

Prior to the tunnel's construction, Carbide engineers extracted core samples from the mountain's interior. Those samples revealed that much of the tunnel would burrow through a vein of pure silica—the essential component required in Carbide's silicon-metals process at Alloy. So valuable was this discovery that the original diameter of the tunnel was expanded considerably; and although the devastating effects of silica dust on human lung tissue had long been chronicled, Carbide officials chose to minimize, or simply overlook, this detail. Essentially the potential dangers related to working in a silica-rich environment were withheld from tunnel workers and, within a matter of weeks, the laborers were succumbing to the fatal effects of silica dust. By the time the tunnel was completed it's estimated that several thousand men were exposed to, and untold hundreds upon hundreds died from, this exposure.

In 1936 a U.S. Congressional Subcommittee on Labor convened to consider allegations related to work conditions in the Hawks Nest tunnel, citing the following explanation for the need of this investigation:

January 16, 1936
HOUSE OF REPRESENTATIVES
[H. J. Res. 449, 74th Cong., 2d session.]
COMMITTEE ON LABOR, Washington, D. C.

Whereas four hundred and seventy-six tunnel workers employed by the Rinehart and Dennis Company, contractors for the New Kanawha Power Company, subsidiary of the Union Carbide and Carbon Company, have from time to time died from silicosis contracted while employed in digging out a tunnel at Gauley Bridge, West Virginia; and

Whereas one thousand five hundred workers are now suffering from silicosis contracted while employed in the construction of said tunnel at Gauley Bridge, West Virginia; and

Whereas one hundred and sixty-nine of said workers were buried in a field at Summersville, West Virginia, with cornstalks as their only gravestones and with no other means of identification; and

Whereas silicosis is a lung disease caused by breathing silicate dust, this dust causing the growth of fibrous tissues in the lung gradually choking the air cells in the lung and bringing about certain death; and

---

8. In 1931, Boncar was renamed "Alloy" to commemorate Carbide's silicon metals alloy production.

> *Whereas this condition has existed for years and all efforts to expose it have been thwarted; and*
>
> *Whereas there are other similar conditions existing in the United States in said industry: Therefore be it*
>
> *Resolved by the Senate and House of Representatives of the United States of America in Congress assembled, that the Secretary of Labor shall immediately appoint a board of inquiry to make a prompt and thorough investigation of all facts relating to health conditions of workers employed in the construction and maintenance of public utilities.*[9]

Following their investigation the committee concluded that Hawks Nest represented a "tragedy worthy of … Victor Hugo." But rather than a fictitious rendering, the tragedy was an all-too-real Armageddon penned at the hands of the Union Carbide and Carbon Corporation, Carbide's quasi-public utility company—The New-Kanawha Power Company—and its contractor, the Rinehart and Dennis Company of Charlottesville, Virginia.

Over the past eighty years several authors have managed to shed light on this tragic event, yet still today a void exists in relation to the accessibility of first-hand documentation relevant to the tunnel's history. This is not surprising for prior to, during and upon completion of the tunnel—as reported in the *Fayette Tribune* in 1931[10] and later confirmed by other reliable sources—Union Carbide issued serious death threats and gag orders prohibiting their employees from discussing the tunnel's construction *problems*. As recently as the mid-1970s, Dr. Lisle Blackwell, a WV Tech professor and noted local historian, halted his own study of the Hawks Nest incident when death threats on his life rose into the double digits.[11]

In 1941 Hubert Skidmore published the first Hawks Nest expose in a novel entitled *Hawk's Nest*. Although theoretically fictitious, *Hawk's Nest* so closely mirrored actual people and events that within weeks the book's publisher, Doubleday, acquiescing to the strong arm of Union Carbide, removed all books from circulation. Coincidentally, on the heels of that purge, Hubert Skidmore died in a mysterious house fire.

Thirty-five years later, Alicia Tyler's article, "Dust to Dust,"[12] was published in *The Washington Monthly*; and finally in 1986, Dr. Martin Cherniack published a stellar analysis of the history, *The Hawk's Nest Incident: America's Worst Industrial Disaster*, noting, "It is remarkable that an event so apparently rich in drama, moral significance, and popular attraction should leave no enduring imprint in popular culture."[13]

---

9. "Investigation relating to health conditions." p. 1.
10. Refer to Chapter Five article dated May 20, 1931.
11. "For Survivors of the Hawks Nest Tunnel, It was Only a Job." *Gazette-Mail Metro East*, July 31, 1996.
12. See Chapter Two.
13. Cherniack, p. 87.

Although other articles have appeared in publications and newspapers, for all practical purposes these three works represent the bulk of readily accessible information related to the tunnel's history.

As a compilation of first-hand accounts, *The Hawks Nest Tunnel: An Unabridged History* opens the tunnel's history for inspection by providing information gathered from a wide variety of sources including congressional testimonies, interviews with tunnel survivors, literary works spawned by the tunnel's events, articles from local and regional newspapers published between 1928 and 1937, documents from the West Virginia Public Service Commission and correspondences from the Federal Power Commission.

Whether digested *in toto* or in part, *The Hawks Nest Tunnel: An Unabridged History* provides ample fare for understanding the enormous social, economic, cultural and political impact of this tragic event, and in so doing will hopefully mend the heart of at least one tunnel worker's relative who wrote, "I drove up there on a cold rainy Friday and stood on the overlook and looked down on the tunnel adit. It was about the loneliest feeling I ever had. All those men who had died putting that thing under the mountain and not one word was written to tell me that a single soul was sorry that it had happened."[14]

---

14. From personal papers of the late Ruby Winebrenner, Gauley Bridge, WV.

CHAPTER ONE

A CONGRESSIONAL INVESTIGATION OF THE HAWKS NEST TUNNEL DISASTER

As a result of extensive regional and national media coverage surrounding rumored travesties at Hawks Nest, Congressional leaders convened in January 1936 to conduct an investigation into the tunnel's construction.

For a nation such as ours—forged by the heat of revolution—it is ironic that individuals who possess the courage to speak out against wrong-doings are often viewed as troublesome, problematic agitators. There can be little doubt that during the 1930s, this would have been especially true. Nevertheless, as the following chapter will reveal, courageous individuals stepped forth to speak out against the outrageous labor practices that allowed the tragedy at Hawks Nest to occur.[15]

The first person to testify was Philippa Allen, a graduate of Brookwood Labor College in Katonah, New York. During the four summers prior to the investigation, Allen lived in West Virginia and gathered information related to the tunnel's construction and its subsequent impact upon the laborers. Articles she wrote, scribed under the pen name of Bernard Allen, were also published in *New Masses*.

Leonard Davis, vice-president of the New Kanawha Power Company, offered the following description of Ms. Allen:

> *Philippa Allen has spent time on several occasions in West Virginia, in the vicinity of Cedar Grove, Pratt, and the Paint Creek region. Most recently she spent some time about Gauley Bridge.*
>
> *At Hollygrove, a mining community on Paint Creek, Miss Allen was associated with people, including one E.J. Sowares, who are remarked for radical proclivities.*
>
> *It is said by people of the Paint Creek valley that Miss Allen is a graduate of Brookwood College, in New York State, which they say is a labor college, and that students of that college have been conducting field operations, in radicalism, in this valley since 1931. They had a community house in East Bank at and from which they conducted propaganda in the surrounding collieries. In this work Miss Allen seems to have been the assistant, or junior to Agnes Saler.*

---

15. Photographs of the men and women who testified during the Congressional hearing can be viewed at **http://corbis.com**, and accessed by entering "silicosis" as the search word.

One resident of East Bank claims to have seen a letter from Professor Rexford Tugwell [16] *to Miss Allen offering her a position in rehabilitation work in the Virginia mountains in the summer of 1935.*

Miss Allen helped organize, and participated in, a march on the governor of West Virginia in an attempt to coerce him in matters of school legislation. Miss Allen and some others who have been at work in the valley are said to belong to the "Pioneer Youth Club." [17]

The Investigation Relating to Health Conditions of Workers Employed in the Construction and Maintenance of Public Utilities convened on Thursday, January 16, 1936. Committee members included Glenn Griswold, Democrat, Indiana, chair; W.P. Lambertson, Republican, Kansas; Vito Marcantonio, Republican, New York; Matthew Dunn, Democrat, Pennsylvania; and Jennings Randolph, Democrat, West Virginia.

Philippa Allen was the first witness.

STATEMENT OF PHILIPPA ALLEN[18]

Mr. Griswold: The subcommittee, which is composed of Mr. Griswold, Mr. Dunn, Mr. Randolph, Mr. Lambertson, and Mr. Marcantonio,[19] will please be in order.

We have with us this morning, to be heard first, Miss Philippa Allen, of the Jacob A. Reis Neighborhood House, 48 Henry Street, New York City. Miss Allen, the subcommittee should like to have you make a statement, if you have one, concerning conditions at or near Gauley Bridge, W. Va.

Miss Allen: As I understand, I am first to tell you what sort of work I am engaged in. I am a social worker connected with the Jacob A. Reis Neighborhood House, New York City.

I have spent the last four summers in West Virginia; and during the summer of 1934, when I was doing social work down there, I first heard of what we were pleased to call the Gauley tunnel tragedy, which involves about 2,000 men.

According to the estimates of contractors, 2,000 men were employed there over a period of about 2 years in drilling 3.75 miles of tunnel to divert water from New River to a hydroelectric plant at Gauley Junction. The rock through which the workmen were boring was of high silica content. In tunnel no. 1 it ran from 97 to 99 percent pure silica, and the contractors neglected to provide the workmen with any sort of safety device.

---

16. Rexford Tugwell, a member of FDR's "Brain Trust," was instrumental in crafting the Agricultural Adjustment Act and the National Industrial Recovery Act during the Depression. In 1935 his duties with the FDR administration included developing greenbelt communities—quasi-utopian urban projects that sought to construct self-sufficient cities from the ground up.
17. Located in the Union Carbide papers located in the Cultural Center, Charleston, WV.
18. "An investigation . . . ." pgs. 2-16.
19. See Biographical Information, page 185.

None of the workmen, some of who have lived around Gauley Bridge all of their lives, were aware of the risk they were running, despite the fact that sandstone outcroppings can be seen all over the roads. These were robust, hard-muscled workmen, and yet many of them began dying almost as soon as the work on the tunnel started. With every breath they were breathing a massive dose of silica dust. That was the true explanation of the deaths.

It usually takes from 10 to 20 years for this condition to develop fully in a man's lungs, but the medical men said that these men were working under extremely dusty conditions and the doses they received were massive indeed.

Silica dust is deadly in large doses. Every worker examined by a physician after working in the tunnel any length of time has been found to have this dreadful disease. It is a lung disease that cannot be arrested, once it is started. Ultimately, the victim strangles to death.

When I tried to tabulate the number of workmen who had died as a result of this condition, I found it impossible to do so for several reasons: First, because before it was generally known what was really killing these men, company doctors had diagnosed the numerous deaths as pneumonia, to which silicosis-infected lungs are susceptible. Second, the undertaker who handled many of the burials testified in court that his records had been destroyed. Third, after suits were started and everybody knew that rock dust was causing this dreadful state of things and killing the men on the tunnel job, workmen left their jobs there and scattered all over the country.

This tunnel is part of a huge water-power project, which began in the latter part of 1929 under the direction of the New Kanawha Power Co., a subsidiary of the Union Carbide & Carbon Co. That company was licensed by the State of West Virginia Power Commission to develop power for public sale, and ostensibly it was to do that. In reality, it was formed to sell all the power to the Electro-Metallurgical Co., a subsidiary of the Union Carbide & Carbon Co., which was by an act of the State legislature allowed to buy up the New Kanawha Power Co. in 1933.

I should like to state that I am now making a very general statement as a beginning. There are many points that I should like to develop later, but I shall try to give you a general history of this condition first.

I found when I went to Gauley Bridge that men were still dying like flies in 1934. These were men who characterized themselves as generally following the mines as a trade. Mining in West Virginia is unsteady, and these men went into this tunnel work because they thought it offered opportunity for steady work at better wages, and that it was work which did not posses the hazards they had met in mining coal, such hazards being poisonous gases and falling rocks.

Of the 2,000 men employed there over a period of nearly 3 years, many have been examined by private doctors. Men began to succumb to the bad conditions within one, two or three years after they started to engage in the work. It seems that only a few of the 2,000 men affected will escape.

Nobody knew of the dangers of this dusty tunnel until the $6,000,000 of lawsuits against the New Kanawha Power Co. and Rinehart & Dennis, contractors, of Charlottesville, Va., were filed. All the lawsuits charged that men were dead or were dying because of working in the tunnel.

The first suits were brought to trial in the spring of 1933. The lawyers representing 300 men settled out of court after the first suit resulted in a hung jury. They settled for a sum of $130,000. The lawyers had taken the cases on a 50% contingency basis, therefore only a very small sum was left to be divided between a large number of men after the lawyers' fees had been subtracted; but not all of the 300 men who were made to sign releases of claims for damages against the power company and Rinehart & Dennis, contractors, before settlement was made shared in the division of the money.

[one paragraph omitted, page 4]

I want to recite some of the conditions that were uncovered concerning working conditions in the tunnel at the trial. The dust was so thick in the tunnel that the atmosphere resembled a patch of dense fog. It was estimated on the witness stand in the little courtroom at Fayetteville, where suits against the builders of the tunnel were tried, that workmen in the tunnel could see only 10 to 15 feet ahead of them at times.

Man after man testified: drillers, drill helpers, nippers, dinkey runners, and members of the surveying crew. As the plaintiff's witnesses, they all told of this dusty condition. They said that although the tunnel was thoroughly lighted, the dinkey engine ran into cars on the track because the brakeman and dinkey runner could not see them. Laird King drove his dinkey into the little one and wrecked it, and Otis Edna, his brakeman, jumped off the front end just in time to save his life. Nippers who took charge of the steel bits could not see the signs given by the drillers when they needed "steel" and the signals had to be relayed. Dust got in the men's hair, on their faces, in their eyebrows; their clothing was thick with it. Raymond Johnson described how men blew dust off themselves with compressed air in the tunnel; if they did not they came out of the tunnel white, he said. One worker told how dust settled on top of the drinking water, "so I took milk in the tunnel with me and drank it instead."

What caused this dusty condition? The use of dry drills, said the workmen. J.J. Huffman told the court how he asked the foreman if a little water could not be used in the hole when the bit got hung up, and the foreman's reply was, "Hell, no."

Milledge Venson said that the foreman stopped the dry drilling while the mine-safety inspector was in the tunnel; and Sam Butner testified that he was stationed at the scaling tower—which was about 600 feet—from the heading, and directed to hurry information to the heading foreman of the approach of the mine inspector so that the dry drilling could be stopped before the inspector reached there. Not only Sam, but Laird King and others told how they had acted as lookouts and warned the foreman when they saw the inspector coming.

Rinehart & Dennis, builders of the tunnel, tried in vain to deny that the workmen

Tunnel under construction. View of Heading No. 2, showing drilling crew. *Courtesy West Virginia State Archives.*

were forced to drill "dry" holes. Albert Young, a Negro worker, originally testified for the contractors saying that there was no dust and that drills were operated by water, but later he appeared in court as a "plaintiff's witness," a witness for the man who was suing, and changed his story. He said there was "considerable dust" in the tunnel and that drills were operated when they were dry, he said. He had been praying since he gave the first testimony and now wished to tell the truth. Before he told his story the first time, he said, he was promised a job and pay by an official of the contracting company if he would testify for the company, and "threatened with the penitentiary" if he did not do so.

Another witness for the contractors was Robert M. Lambie, former chief of the state mines department, who said that the tunnel was practically dust-free when he made inspections in 1930 and 1931. He told the jury that the men were easily distinguishable from 500 to 700 feet away, and that drills were operated with water.

Why did he say this now, when in 1931 he had written letters to the contracting company instructing them to remedy the dusty conditions in the tunnel, the plaintiff's lawyer asked him. Lambie said he had been misinformed by his inspectors concerning conditions in the tunnel in 1931. He admitted that he had recommended the use of masks and respirators at the time; but he said later he withdrew this recommendation after a conference with the contractors, when he decided that masks were not necessary.

Throughout the court trials the witnesses for the contractors gave the flimsiest testimony. O.M. Jones, chief engineer of the New Kanawha Power Co., "never saw dust, or at least enough to say it was dusty." He saw fog and mist in the tunnel; but "the air was as clear as it was in the courtroom, except on foggy days."

Under cross-examination, Owen Jones admitted that he had received a letter from Lambie, mine safety inspector, on May 18, 1931, saying that the heavy concentration of silica dust in the tunnel was highly dangerous and giving orders that respirators be used by the workmen.

The contractors tried to show in court that they had not been negligent in making arrangements to care for the safety of men on a construction job of this sort. Engineers from other contracting companies were called to testify that their companies made a practice of drilling "dry"; that respirators were not necessary; but, regardless of the legal facts, many hundreds of West Virginia miners who contracted to push the 3¾-mile bore through the mountain paid for their jobs with their lives.

Last year the first 167 lawsuits filed against the Rinehart & Dennis Co. were settled out of court when a payment of $130,000 was made to the lawyers of these 167 men suing; but there still are more than 200 damage suits that were brought by workmen or workmen's widows against the contractors which have not been settled. They will come to trial later.

It came out in court proceedings that the company doctors were not allowed to tell the men what their trouble was. A Dr. Mitchell, of Mount Hope, a company doctor, testified for the plaintiffs, saying that he had told the men they had "tunnelitis."

It was a Mrs. Jones who first discovered what was killing these tunnel workers. Mrs. Jones had three sons—Shirley, aged 17; Oren, aged 21; and Cecil, aged 23—who worked in the tunnel with their father. Before they went to work in the tunnel, Mr. Jones and Cecil and Owen worked in a coal mine; but it was not steady work, because the mines were not going much of the time.

Mr. Griswold: Can you give us the Christian name and address of Mrs. Jones?
Miss Allen: Mrs. Charles Jones, Gamoca, W. Va.

Then one of the foremen of the New Kanawha Power Co. learned that the Joneses made home brew, and he formed a habit of dropping in evenings to drink it. It was he who persuaded the boys and their fathers to give up their jobs in the coal mine and take on this other work, which would pay them better. Shirley, the youngest son and his mother's favorite, went into the tunnel, too.

Mrs. Jones began to be suspicious when she saw the amount of sediment that was left on the bottom of the tub after she had washed the clothes of her men folk. She asked the foreman about the dust, and he said it was just ordinary dust and would not hurt anybody. Then one day Shirley came home and complained, "Ma, I'm awful

Lower Gamoca, March 4, 1936. *Courtesy West Virginia State Archives.*

short-winded." She said to him, "Well, if you never feel no better, you'll not work no more." That was in September of 1931, and he died in June 1932.

[paragraph omitted, page 6]

She told us that the boy's last wish was that, "Mother, after I'm dead, have them open me up and see if I didn't die from the job. If I did, take the compensation money and buy yourself a little home." Within 13 months of Shirley's death, Cecil and Oren also died.

Shirley Jones' case was the first of a long line of lawsuits to be filed against the Rinehart & Dennis Co., of Charlottesville, Va., contractors to whom the New Kanawha Power Co. had allotted the work of drilling the tunnel.

Although the workers testified that the dust was thick in the tunnel headings the company officials dared to deny it. How many workers do you think took the company's side when called to the witness stand? One foreman, who has since died of silicosis, and one colored worker, who later changed his story because his conscience bothered him. Two men too many, don't you think?

When suit was instituted in Fayette County, W. Va., it was decided, on appeal, that silicosis was not compensable from the State compensation fund, and that Mrs. Jones was therefore entitled to sue. That applied to all other victims and their kind.

When I first went up to Gauley Bridge, Harless Gibson told me to find out about the whole situation of the tunnel work. People would not believe it. He told me, "You look at that tunnel and you think it is a fine thing when you do not know how many men died in building it. You cannot say anything too bad about the tunnel work; but people will not believe it."

Living conditions of the men were as bad as the conditions of their work. As high as 25 to 30 Negroes used to sleep in a shack no larger than 10 by 12 feet. They were made of Jerryline stripping with a half window in the side and a home-made door. There were two bunks stretched across the side of the room, and he said, "I have observed as many as 15 men piled in a heap on the bunk."

When we said we would like to talk to some of the men who had worked in the tunnel, Mr. Gibson called to a colored man who was passing, "Come here, George, and tell these ladies your story." And George Houston, a hard-muscled, strongly built man of 23, came up to us walking very slowly and breathing with effort. He is in what the doctors call "the third stage" of silicosis, which means that he has not much longer to live. There were dark rings under his red-rimmed eyes, and when he climbs stairs, "It gets me to breathing so hard I have to lay down," he said. George worked only 48 weeks toting water, shoveling muck, or operating a drill in no. 1 heading of the Gauley Junction-Hawks Nest tunnel, yet in that short time he breathed so much silica dust in the badly ventilated heading that the disease is rapidly destroying his lungs. We asked George how much rent he had paid for sleeping space in one of the box-like hovels Mr. Gibson had described, and this is what he said:

Fifty cents a quarter (a week). They furnished only a little old shack. We paid shack rent every Friday. There was nothing in the shack. The men had to buy bedclothes, coal, and a stove. They used to bring the old dynamite boxes up from the workings to set on. Men, women, and children were crowded up together. Some of the women were married, and some wasn't. Families had four, five other men sleeping on the bunks with them. Some men couldn't stand the conditions of the shacks. You could see they was lousy if you looked in. I went to stay at the Jungle, at Gauley Junction between the railroad and the river, but I had to pay to stay over there. They took shack rent from anyone who had a working ticket."

The well went dry. "They had a tremendous spring up at the camp which dried up," George said. "Then we had to walk 2 miles for water."

How much was coal? "At first we had to pay 25 cents a quarter for coal, then they raised it to 50 cents. Every year they raised it." Each man who worked paid for coal, whether or not he used it.

Wages were cut from the 50 cents an hour that men were paid in 1930, to 40 cents and then to 30 cents an hour.

They made the men work whether they wanted to or not. They made the men work if they were sick or not. They had a shack rouster named McCloud, who carried a gun. He was a deputy sheriff licensed by Fayette County, the license having been given on recommendation of the New Kanawha Power Co., and every morning he went up to the shacks and made the men to go work (sic). McCloud threatened to jail men who would not work. When George's partner in the drill had his head cut off by falling rock, George did not want to go back into the tunnel, therefore, Deputy Sheriff McCloud arrested him.

Mr. Randolph:[20] How many hours a day do those men work?

Miss Allen: Ten hours for each shift.

Mr. Marcantonio: Who is Gibson, of whom you have been talking?

Miss Allen: Harless Gibson was then what they call down there a runner for a firm of lawyers of the name of Townsend, Bock & Moore, and he was taking care of their cases. From 1928 to 1932 he was deputy sheriff of Fayette County, W. Va., as I am reliably informed.

Mr. Gibson told us more of this "shack rouster." In No. 1 camp for colored people McCloud ran a club for men, a place where they could drink and gamble. "It was a skin game," Mr. Gibson said. "The cut" for the house was 25 cents when betting on cards. He chased the "niggers" in from the hills if he found them throwing dice and made them gamble in the clubhouse so that he would get a percentage of their winnings. He would take all their money away from them and give it back to the company.

What had become of McCloud, we asked. The religious people in Gauley Bridge complained about the gambling, and C.A. Conley, head sheriff of Fayette County, after he had warned McCloud several times to stop the gambling, went up and closed the club house. He took McCloud's commission away from him at the same time. McCloud, his usefulness to the New Kanawha Power Co. being over, is now trying to get a job with the Koppers Coal Co.

The majority of the men working on the tunnel, who died when the work was first started, were colored men. Perhaps because negroes catch lung diseases more easily than white men do. Mrs. Jones, who lives at Gauley River, told us that, "They buried them like they were burying hogs, putting two or three of them in a hole. The men were buried ill what they got killed or died in."

The story of the treatment of the colored men on the job at the tunnel is the same old one of racial discrimination. Look at how much they were docked each week for the company doctor, 75 cents, which was 25 cents more than the white workers paid. They paid 75 cents weekly for the services of a doctor who never came to see them. "I sent a call for the doctor for 4 weeks and he never came," George Houston said, "and I was still paying for him."

We heard of instance after instance of brutal treatment and discrimination. "They was treated worsen than if they was mules," Mrs. Jones told us. "The foreman would cuss at

---

20. Jennings Randolph served in the U.S. Congress as a representative from West Virginia's 2nd district from 1933-1947. From 1958-1985 he represented West Virginia as a United States Senator.

East dam cut-off wall construction; excavation beyond the C&O Railway tracks. *Courtesy West Virginia State Archives.*

them bad and run them ragged. He would run them right back into the powder smoke in the tunnel after a shot, instead of letting them wait 30 minutes like the white men do."

Why did the Rinehart & Dennis Co., contractors, dare to treat the colored "worsen if they was mules"? Simply because these poor, ignorant men had no standing in the community and there was no friendly organization to which they could protest. Most of them were far from home. They had come in droves from states up and down the Atlantic seaboard, from Pennsylvania, Georgia, North Carolina, South Carolina, Florida, and from states as far inland as Alabama, Kentucky, and Ohio.

Most of them had been recruited by scouts of the company who went through the States giving glowing accounts of "steady work" in Fayette County. In this way, a steady stream of cheap labor kept pouring in, enabling the company to reduce the hourly wage until it reached a low of 25 cents. Unorganized workers everywhere are an (sic) easy prey for the large companies who rob working men.

This greedy company of contractors, Rinehart & Dennis, not only robbed its workers by a ridiculously low wage scale, but purposely doomed them to die when they neglected to furnish men respirators (masks), which would have kept them from inhaling the deadly silica dust in the tunnel headings.

Kies, purchasing agent for Rinehart & Dennis, was overheard to say to a respirator salesman, "I wouldn't give $2.50 for all the niggers on the job." Kies was voicing the hatred and greed of this large company for which he worked. Ernest Lyes, a white man of 26, testified in court that he heard Kies say this. I have an affidavit covering this matter.

Why do you think the contractors from Charlottesville, Virginia, dared not furnish their workers with safeguards of masks and wet drills? Because they thought they would finish the job and be out of the state before the men began to die. Silicosis usually takes from 10 to 20 years to develop in one's lungs. Kies spoke again for the company when he said to Hawkins, the assistant superintendent, "I knew they was going to kill these niggers within 5 years, but I didn't know they was going to kill them so quick." George Houston made an affidavit saying he heard Kies say this in the company commissary where George had gone to buy a can of tomatoes.

Almost as soon as work was begun in the tunnel the colored men began to die like flies, because the percentage of silica in the dust they inhaled was so large.

The ambulance was going day and night to the Coal Valley Hospital.[21] As soon as a man died they would bury him, we were told. One colored boy died at 4 o'clock in the afternoon and he was buried at 5 o'clock the same afternoon without being washed. Why? Because the company did not wish an autopsy performed, which would have uncovered the cause of his death.

If word of the terrible disease killing the men had reached their ears, do you think they would have stayed on the job? The tunnel needed to be finished quick, quick, quick. Profits were all that interested the company.

Mr. Dunn: Can those statements be verified; can we get those facts; can it be shown authoritatively that the colored boy died at 4 o'clock and was buried an hour later without being washed?

Miss Allen: Most of this I am saying is court testimony. It has been very hard to trace these cases.

I have one story that, I think, is outrageous. Harless Gibson, who was a deputy sheriff, told me how a woman had taken out a warrant against an undertaker to obtain the body of her husband. That was from the head sheriff, C.A. Conley, of Fayette County. A process server tried to get service on him at the hospital in Montgomery, but he escaped into Nicholas County, where he had a burial ground. Conley had to call the head sheriff of Nicholas County and get him to serve a warrant on the undertaker before the body could be obtained.

H.C. White, a Summersville undertaker, was given a contract by the company to bury the Negroes at $55 a head. He buried the men on his mother's farm outside of

---

21. Coal Valley Hospital later became Laird Memorial Hospital. Today it's known as the Montgomery General Hospital.

Summersville, and it is rumored that this plot of ground was plowed and planted with corn. This fee of $55 a head for burying was lower than other undertakers charged.

Mr. Randolph: Did they not advertise for bids in connection with this work?

Miss Allen: No. I was reliably informed that the undertakers at Gauley Bridge would not bury them, but White was hired by the contractors to bury them at $55 a head. The men charged in their suits that the reason he took the work for a smaller price than the local undertakers charged was that the company knew and assured him there would be a large number of deaths.

Mr. Griswold: What does a county or a municipality pay in that community for the burial of a pauper?

Mr. Randolph: $30 each.

Mr. Griswold: Which is $25 less than White received?

Mr. Randolph: Yes.

[paragraphs omitted from end of page 10 to bottom of page 11]

Miss Allen: I have visited with a friend the little construction camp, which is high in the hills, and we found that many of the shacks had been torn down. The company had torn them down to get rid of the workmen who were suing. One man said they had threatened him with jail, but they would not run him out before he died or they paid him.

These simple people crowded around us and asked, "What are you going to do to help us; what are you going to do?" They said the same thing again this year when I went back, and I found that the cases were being settled, some being paid and others not. All were bewildered. Again at Vanetta they are asking, "What can be done about this; won't you please help us?"

I feel that this investigation may help in some manner. I do hope it may.

Mr. Dunn: We are willing to make an effort to help those unfortunates.

Miss Allen: I should like to go more specifically into my material, telling stories of a man and checking up statements about the dust in the tunnel, giving concrete statements.

I returned to New York City with the cry, "What are you going to do for us?" ringing in my ears.

I attended the annual meeting of the stockholders of the Union Carbide & Carbon Corporation, of which the New Kanawha Power Co. is a part. I had a proxy from one of the shareholders. She wrote me a letter saying that the stockholders ought to be told how many men were dying to make money for them. My chief hope was that the press would carry this story so that it would be well circulated all over the country; but, unfortunately, none of the papers carried it despite the fact I held the floor for a half an hour asking questions.

Mr. Randolph: That was a stockholders' annual meeting of the Union Carbide & Carbon Co.

Miss Allen: Yes. The meeting was held on or about April 16, 1935. This is a little introduction that amused me. It tells how I happened to bring in the first question.

Many of the shareholders at the meeting were nervous about the divisions of the profits, when it came time to discuss next year's plans affecting them. A special compensation plan for employees was discussed. It would take not more than 7.5 percent of the company's net income. Mr. Wood protested, "Is the company forced to take this measure?" The gist of his remarks was that "I prefer to be an employee. Employees never take the losing side." The plan proposed was overly paternalistic. He meant that the company was too good to its employees.

Mr. Jesse J. Ricks, the president of the Union Carbide & Carbon Corporation, suggested that the stockholder had better take this question up in a private conference. Therefore Mr. Wood withdrew his question.

But I spoke up. I said this was important. "Has the company been forced to draw up plans for compensation?" "No," said Mr. Ricks in an unconvincing tone.

"Then how much has the company spent on lawsuits for workers dying of silicosis at the Gauley Tunnel in West Virginia?" I asked. President Ricks did not know, but a man sitting on his left volunteered the information. He said $150,000. Then a small man with a wild glare bounced out of his seat to stand over me. "I am familiar with this case", he said grimly. "We haven't spent a cent." He was Attorney Smith, special counsel for West Virginia matters. When I asked Mr. Ricks why the $150,000 had been spent he replied, "Oh for his expenses and salary," pointing to Smith.

Mr. Smith said that the men who worked in the tunnel were employees of the contractors, Rinehart & Dennis, only. I read him the terms of the contract made by the New Kanawha Power Co. with the contractors, pointing out that the chief engineer of the power company was not only given the power and authority to direct any change in work in the tunnel, but also the right to hire and discharge any men he wished. "Doesn't this make the New Kanawha Power Co. responsible for the acts of the contractor under the master-and-servant law?" I asked. Of course, that law makes the master liable for injuries done to servants who are carrying out his orders. There was no reply to my query. For further information I was told I could talk the matter over privately with Attorney Smith. Thus a great corporation silently disowned those who made its wealth.

Mr. Randolph: The Union Carbide & Carbon Co. has holdings in many States, has it not?

Miss Allen: Yes, it has. It is one of the largest, if not the largest—if its rating has not changed recently—concern of its kind in the United States.

Mr. Griswold: Is Mr. Smith located in New York City or in West Virginia?

Miss Allen: He has an office in the Union Carbide & Carbon Co. Building in New York City.

Mr. Griswold: Have you the excerpts about which you just spoke?

Miss Allen: Yes.

Mr. Griswold: Will you please enter them in the record, or give them to the reporter and he will do so.

Miss Allen: Quoting from specifications and contract of Rinehart & Dennis and the New Kanawha Co. for the construction of this tunnel, I find in articles X and XX, in part, the following:

> *The contractor shall take all responsibility of the work, and take all precautions for preventing injuries to persons and property in or about the work; . . . In any case where, in the opinion of the engineer, injuries to any person or corporation or damages to any property are likely to result from any acts or negligence of the contractor, or any of its agents or employees, the engineer shall have the right to employ such measures as he may deem necessary or desirable to effect a satisfactory avoidance of such injuries or damages, and, if, in his opinion, the case appears urgent, he may proceed to employ such measures without previous notice to the contractor, which, however, shall not be relieved from any responsibility on account of such action of the engineer . . .*
>
> *Specifications: section 124. Ventilation: The contractor shall keep the tunnel air in a condition suitable for the health of the men, and clear enough for the surveying operations of the engineers. All possible precautions shall be taken to keep dust from drilling within such limits as will not be injurious to health.*

A sufficient supply of fresh air shall be provided at all times in all places underground, and provisions shall be made for the quick removal of gases and dust generated by blasting, or by dust-producing if any be installed in the tunnel. Ventilating plants, of ample capacity, shall be installed and used (until and unless rendered unnecessary by natural ventilation after headings meet) while work is going on in the tunnel."

There was not a word in the press about this condition, despite the fact that a World Telegram man before the conference introduced himself to me and informed me that he was interested in what was going to follow.

I went up to interview this lawyer two days later, taking a friend of mine with me to check any statement that might be made; and I think you would be interested in one thing he told us. Of course, he denied liability of the company, the New Kanawha Power Co., but he said that he had advised the contractors that they were not liable, he arguing that the men had not had a physical examination and they had not contracted silicosis on the job, despite the fact that Mr. Ricks himself only 2 days before had answered my question in the affirmative, supposedly that they had made some settlement with the men who were dying of silicosis in the Hawks Nest Tunnel when he said $150,000.

Mr. Ricks said liability was with the contractors, and then we asked whether he was covered by insurance in the event the contractors were unable or failed to fulfill their contract on time, for instance. He said, "Yes; all big companies are so protected." He said the Union Carbide & Carbon Co. was covered by $4,000,000 bond held by two surety companies to protect itself against defaults or other liabilities of the contractor. Then he seemed to be very much upset by what he had said. He showed tremendous distress and continued to say, "This is not important." He talked about it so much and so earnestly that the psychological effect on us was quite contrary to his statement of its unimportance. If we had been wrong, he would have dropped the subject soon, no doubt.

He had been talking about general liability, and we asked him if that did not mean labor liabilities, and he said it had not been interpreted as such in the courts as yet. That was in fact the only new thing he told us.

I went to West Virginia in the summer of 1935 and one of the first things I did when I reached there was to go to the law offices of Townsend, Bock & Moore, which firm had helped me in the summer of 1934 in gathering quite a lot of material. While waiting to be received by a member of that law firm I sat down beside a Negro worker who was the victim of silicosis and the wife of another worker who was too sick to come himself. The Negro muttered aloud angrily, impatient at the delay. This workman said aloud to himself with much bitterness, "You wait, wait, wait, and the boss man gives the poor man nothing. . . . He ain't ever goin' to . . . The poor man kaint git nothing." With that wail ringing in my ears I was called in to interview the attorneys of this victim and many others. I asked many questions. "How many cases of the men who sued did the settlement cover?" "Which company made settlement, the power company or the contractors?" "What Charleston lawyers represented the companies in the settlement?" All the while the lawyers sat mysteriously silent, smiling like sphinxes. Ben Moore after a space said, weighing each word as though it were of tremendous import:

> *We are not at liberty to answer your questions. In this case there is a certain professional obligation to the other side not to disclose any facts they might not want given out. All we can say—yes; I think we can say—is that the settlement was comparatively small.*

To all my questions the attorneys replied evasively "it is not known," or "we do not know this." "How much was the settlement?"

> *"The men will know when they are paid," volunteered Attorney Bock.*

But men who brought suit, with whom I talked recently, are still wondering, waiting to be paid off. One man told me of a widow of a Negro workman who was given $85, her share as determined by her lawyers. I asked this man, Howard McAttee, a literate white man, how much he had sued for, and he said, "I don't know." When I asked Attorney Bock this same question he returned the same answer. Perhaps the company's attorney will know, I thought with some anger, but again when I asked for the name of the company's attorney Bock and Moore were cagey and said:

"We cannot tell you that. We have a professional obligation to the other side not to divulge facts they might not wish to have known."

Veneta [sic] is a small town 2 miles from Gauley Bridge. It is an abandoned town.

It used to be filled with men when the tunnel was underway, but there are only a few who linger there. I have seen there recently about 100 persons in all. I do not know how many families are there; something like 34, perhaps.

I wonder whether you would like to see photographs of the workers in Vaneta and the silicosis victims? I have here [indicating] such photographs.

Mr. Griswold: Yes. You may pass them around to members of the subcommittee as you continue your presentation.

[omission of section, page 16]

Mr. Griswold: Do these [indicating] photographs represent victims of silicosis?
Miss Allen: They do.
Mr. Dunn: About how young are the men who worked in that tunnel?
Miss Allen: Shirley Jones started to work in the tunnel when he was 16 years of age, going on 17. I found men 21 or 23 who had this disease. I talked to them. On the other hand, Clev Montgomery was near 40 years of age. Most of the workers were quite young; younger than 50. I do not know Mr. Jones' age.
Mr. Dunn: You have told us that about 2,000 workers contracted silicosis.
Miss Allen: That is an estimate by the contractors. There was a high turn-over. I asked Mr. Smith, attorney for the Union Carbide & Carbon Co., the largest number of workmen employed at one time in drilling the tunnel, and he said 700. He said, though, that such figure was only an estimate—a guess. There was a large turn-over in the period of a little more than 2 years when they employed 2,000.
Mr. Griswold: Was this large labor turnover of which you speak caused by the men having silicosis or by unfavorable working conditions? When the men were working there they did not know they had silicosis, did they?
Miss Allen: They did not. Men began to leave as soon as word spread that other men were dying on the job.
Mr. Griswold: It was an unhealthy place?
Miss Allen: Yes—and that word spread all over the country. I have heard of New York City hospitals, in connection with post-graduate work, sending down to West Virginia for case histories of those men.

[omission, page 16]

Mr. Griswold: It is time for the House to convene and for member of the subcommittee to go to the House. Therefore, the committee will adjourn for the day, to meet at 10:30 o'clock tomorrow morning.

(Thereupon, at 11:55 a.m., Thursday, Jan. 16, 1936, the subcommittee adjourned, to meet at 10:30 a.m., Friday, Jan. 17, 1936.)

FRIDAY, JANUARY 17, 1936[22]

Mr. Griswold: The committee will come to order. We will resume hearing the testimony of Miss Allen. Miss Allen, will you come forward, please? You are the same Miss Philippa Allen that testified yesterday?

Miss Allen: Yes, sir.

Mr. Griswold: All right; you may proceed.

Miss Allen: In my testimony yesterday I tried to give a general picture of the situation at Gauley Bridge, W. Va., where hundreds of men who dug a power tunnel through silica are dead or dying. Today I would like to show how that general situation affects the men and their families by giving several specific examples. I shall be brief so that you may ask me any questions you wish.

Yesterday I mentioned the Jones family briefly; Mr. and Mrs. Charles Jones and their children, of Gamaca, W. Va., a small community near Gauley Bridge. Five of the men in the family worked in the tunnel—the father, Charles Jones; three sons, Shirley 17, Cecil 23, and Oren 21; and Mrs. Jones' brother, Raymond Johnson. The father and the two older boys were coal miners. Shirley had never worked. The glowing stories of how well the tunnel work was to pay, and how safe it was to be, in comparison to mining, persuaded them to start work for Rinehart & Dennis. Even Shirley was brought in for his first real job.

That was in September 1930. In June 1931 Shirley was ill. In September 1932 he was dead. Within 13 months Cecil and Shirley died, in that order. In November 1934 Raymond Johnson died, after having been bedridden for 6 months.

I hope that you bring Charles Jones to this hearing. His friends are wondering whether he can live until Easter. His breathing shows the unmistakable signs of silicosis. He has lost weight seriously. They are not doctors to diagnose the case.

[omission, top of page 18]

But they know the signs of coming death. He knows the signs, too.

He has three small children still dependent on him, as well as Cecil's widow and a grandchild. He cannot get work; the small amount of compensation money he received is gone; he is too weak for work relief. He just sits there waiting to die. The debate as to how much dust was in the tunnel goes on, with doctors and lawyers and engineers disagreeing about details. But there is the Charley Jones family—one family, four dead, and one dying.

At Vanette [sic], another small community near Gauley Bridge, I talked with Mrs. Thelma Andrews, formerly of Salisbury, N. C. She is the widow of Sidney Andrews. Sidney, who was 27 when he died, had never been sick a day in the 4 years of their marriage when he became ill on the job. He was taken to Coal Valley Hospital and in

---

22. "An investigation…." pg. 17-33.

4 days was dead. It has been claimed by Rinehart & Dennis that they were generous in their settlements, but Mrs. Andrews told me that she had sued twice and that she had never received any settlement for the loss of her husband. She would have been content with a trifle—enough to get her back home to her family. Now she waits, hopelessly, in a community where there is no work she can get, and no way she can get back to their old home. She was one who asked me, "Can't you do something to help us?"

[one paragraph omitted]

Nancy Jones . . . is the widow of Lindsey Jones. Lindsey worked for 9 or 10 months as a drill helper. He was one of the first to die, on June 23, 1932. Nancy told me how she was one of the first to institute suits and finally, in 1935, she received a settlement from the contractor. I asked her how much it was, but at first could not persuade her to tell me. She was ashamed, she said, because it was so little. Finally she confessed they had valued her husband at $185.85.

"I don't know what that 85 cents was for," she added bitterly. She didn't seem to think it was a generous settlement for the death of Lindsey. He died at the Coal Valley Hospital, and the cause of death was marked as pneumonia, as were most of the silicosis deaths at the beginning.

Jake Swetman is another example of those who were brought in from other parts of the South and want to go home before they die. He was from Orangeburg, S. C. He worked 20 months, he said, a part of the time outside the tunnel, during parts of 1931 and 1932. He was drilling, nipping steel, and, mucking. He went to a Charleston doctor in 1933 and was told that he had silicosis in the first stage. He prepared papers to sue, but E. J. Perkins, Rinehart & Dennis' chief engineer, said, "There was enough for everyone to have enough to go home on."

Anyway, the suit was never filed by the lawyers, and Jake never had any compensation. Last summer, when I talked with him, he was still waiting for money to go back to Orangeburg. He was beginning to suffer painfully although he still looked well.

It was Leo Grey, at Vanetta, who first told me of the "little black devils." That sounds like superstition but was just an angry name for the black pills the company doctor gave them whatever happened. He had worked in the tunnel a year in 1931, and in 1932 he became ill. "My head, stomach, and side began to hurt, and I went to Dr. Mitchell," he told me, "but all I got was little black devils. If rock fell on one of us they just gave us those little pills." When I talked to him he was trying to live by picking berries, with what food he could beg from neighbors at nearby Gauley Bridge. He was unable to handle a job that would support him. Nor could he get relief or any compensation from the company. He just sits in Vanetta awaiting death.

[two paragraphs omitted]

Boys and men have died; their families have been left destitute. Other men are now coughing away their lives, unable to support their families and looking forward only to death. These facts cannot be denied, they cannot be covered over by statements that the tunnel had no dust but on the other hand had marvelous safety devices. There stands the record of death and suffering and want. Behind that record other facts stand.

First, the New Kanawha Power Co., Union Carbide & Carbon subsidiary, had geologists who had made test bores and who knew that the tunnel was to go through pure silica. Then they enlarged the tunnel of project No. 2 from 32 to 46 feet at the location of the richest silica deposit. This was to enable them to take out more valuable silica rock, which was loaded on cars at the tunnel mouth and shipped on the C&O tracks down to Alloy, W. Va., plant of the Electro Metallurgical Co., where it was stored in the yard. It was so pure that it was used without refining. Knowing that this was pure silica, these contractors, with 30 years' experience, must have known that there was danger of silicosis for every man who worked in that tunnel. As Attorney Bock, of Townsend, Bock & Moore, Charleston, pointed out at the Donald Shay trial in Fayetteville, "the engineers of the New Kanawha Power Co. used masks when they went daily in the heading gathering samples of rock."

Second, the men did not know of the danger they were being sent into, because E.J. Perkins, superintendent of Rinehart & Dennis, did not post notices of the danger as required by law, so that the workers did not voluntarily assume the risk. Many of the workers came from agricultural communities in the South where the disease was unknown. They were not experienced tunnel men or hard-rock miners who would have known. Their testimony is universal that it was not until the "ambulance was clanging day and night to the Coal Valley Hospital" that they realized there must be something wrong. Then, there were various diagnoses, one doctor finally hitting upon the word "tunnelitis." When the men realized the danger, it was too late.

Third, as Dr. Emery R. Hayhurst,[23] chief of the division of hygiene of the Ohio Department of Health says: "These men need not have died." There are safety devices available that would have saved all or most of the lives lost and those which will be lost.

It is agreed there was a 24-inch ventilation duct[24] in the tunnel either all or part of the time. Some say it was not put in until the state mine-inspection service forced it on the company; and the men tell of the times after it was put into use when it was not functioning. All the men agree that it was totally inadequate, as does Dr. Hayhurst. The men tell how they would go to the mouth of the tube, as they call it, to get a breath of fresh air just like a group of chickens go to a dish of water for a drink. There they could get the effect of the feeble flow of air coming into the tunnel; back a few feet it was lost in the clouds of silica dust. The tube was full of holes, as no one disputed at the trial.

Interesting evidence was given by W.C. Boxley, a contractor who testified for Rinehart & Dennis. He was then driving a tunnel working only 20 men, half the size of the Gauley Bridge tunnel and he used a 24-inch vent tube with a 24-inch fan. In

---
23. See "Partial Testimony of Dr. Emery Hayhurst," Appendices, pgs. 191-202.

Tunnel opening and ventilation pipe. *Courtesy West Virginia State Archives.*

silica, at Gauley Bridge, a 24-inch tube was used and only an 18-inch fan.

Again, the amount of dust could have been cut down by the use of wet drilling throughout. This was one of the most hotly fought points of dispute at the Donald Shay trial, the executives testifying that there was wet drilling and the men who did the drilling testifying that most of it was dry drilling. Charley Jones told me of the men warning the foreman when the State inspectors came so that the dry drilling could be stopped. He did not realize then what was going on before his eyes: that it was going to destroy the lives of his three sons and his own health—the men were just "trying to keep the bosses in the clear," as he explained it to me. Here again is a disputed point where the final deadly results seem to arbitrate the differences in testimony. Only one man who worked in the tunnel, a foreman, testified for the company at the Shay trial—and he was racked with the silicosis cough as he testified.

As to the most important point, the neglected protection of the health of the men, there was no difference of opinion. Clearly the men were not furnished respirators or masks. The men say that some of the engineers wore masks when they came into the tunnel, and one told me that he bought himself a mask when he saw the headmen wearing them. But the 2,000 men who went into that silica tunnel in all ignorance of what it meant to them in the future were given no masks of any kind. There is absolutely no debate on that question.

Two years ago there was much debate as to the amount of silica dust in the tunnel. Now the men who swore there was no dust and those who said there was are impartially the victims of that dust, and it would seem pointless to prove the fact any more effectively than the witnesses' deaths.

In summary, the men did not and could not have known of the danger they underwent. The company did know the danger they were sending these men to face. They deliberately failed to furnish sufficient protection. The results have been devastating in their deadliness.

[omission several lines]

Mr. Randolph: You have made the statement here, Miss Allen, yesterday, that the undertakers' records of burial had been destroyed?

Miss Allen: Yes, sir; I did.

Mr. Randolph: By whom do you believe they were destroyed, or do you have any information as to that?

Miss Allen: No. I have not. That is simply what was said in the testimony at the first trial.

[omission page 22, first paragraph on page 23]

Mr. Randolph: You made the statement yesterday, I believe, that men were dying like flies?

Miss Allen: Yes.

Mr. Randolph: Would you just amplify it a little for me?

Miss Allen: That is what I was told. And what I found out as I talked to man after man. They are rather hopeless; they sit around. You will find them by the roadside and just waiting to die. They know they have it and there is no hope for them.

[omission, part of page 23]

Mr. Randolph: Now, you spoke of the dust being so thick that they could not see more than 10 or 15 feet ahead of them from the tunneling. Was I right in getting that information from you yesterday?

Miss Allen: Certain of the men testified they could not see more than 10 or 15 feet ahead of them, yes.

[omission, part of 23-24]

Mr. Griswold: Mr. Dunn?

Mr. Dunn: Miss Allen, I understood you to say yesterday that you had visited some of the sheds where the men were compelled to live?

Miss Allen: Yes.

Mr. Dunn: Did I understand you to say that they were about 10 by 12 feet, and 15 men were compelled to sleep in them?

Miss Allen: Yes; what I read yesterday, I think the figures show that. Yes; 15 men. The man who told me that was Mr. Gibson, the deputy sheriff of Fayette County, 1928 to 1932, and I looked at them myself and they were very small.

Mr. Dunn: Were there white men sleeping in those camps?

Miss Allen: They had separate camps for the whites and the colored.

[omission, part of page 24]

Mr. Dunn: Do you know how many hours a day they were compelled to work?

Miss Allen: Yes; there were 10-hour shifts, but the men told me they were made to work until they had cleaned up all that they were supposed to do during that day, clean out the muck and have it ready for the next shift coming on. Sometimes it would be 12 hours and more, and they were not paid for overtime.

Mr. Dunn: What was their salary?

Miss Allen: It started out at 40 cents and dropped to 25 cents an hour.

Mr. Dunn: Do you mean that 40 cents an hour was the highest price paid for labor for work done in that tunnel? Was that the highest figure they had given their employees?

Miss Allen: Yes.

Mr. Dunn: Miss Allen, when did they start construction of this tunnel?

Miss Allen: They started it, as I understand it, in 1930. I think a little preparatory work was done in 1929, probably the engineers plotting the course of the tunnel.

Mr. Dunn: Is it completed now?

Miss Allen: The last reports were, I think, that the steel and concrete work has been completed.

Mr. Dunn: How many men were employed in the construction of that tunnel? I believe I understood you to say about 2,000?

Miss Allen: That was just on tunneling. I have the figures here, if you are interested.

Mr. Dunn: Can you give me approximately how many were employed?

Miss Allen: On the whole job, the total number of white men was 1,700, and the total number of Negroes was 3,100.

Mr. Dunn: Out of this number, how many have contracted this disease; what figure has been given to you?

Miss Allen: Well, I would judge from the number of men suing them—that would be the only figure I would have—those suing the company, and that reaches a total of, well, we have 255 suits after the first 300 suits, and that makes a total of 555 suits instituted.

Mr. Dunn: Were there maybe 1,200 or 1,500 that did not sue?

Miss Allen: Yes. The men scattered all over the country and perhaps they do not know what they have.

Mr. Dunn: I understood you to say that these unfortunate men who have contracted that disease are really in need of assistance, and it is very hard for them to get it, and that some of the men are going from door to door, soliciting food?

Miss Allen: Yes.

Mr. Dunn: Because they are unable to work?

Miss Allen: Yes, sir.

Mr. Dunn: What would you say the condition was in that location generally? Is that the condition of practically every man who has contracted that disease?

Miss Allen: Well, the relief situation is pretty bad down there. The relief office is quite far from Gauley Bridge and the men have to walk about 14 miles, and they get down there and find the relief officer cannot take care of them, most of them. The victims share with the ones who are able to work, and I have seen George Huston coming back with a loaf of bread in his arms that he got from a colored friend who was working, making very little and giving what he could to his friends, more than he could afford.

Mr. Dunn: You have met these people personally?

Miss Allen: I have talked to the people; yes.

Mr. Dunn: Were you present in court at any time?

Miss Allen: There were no cases being tried.

Mr. Dunn: Are there any cases pending now?

Miss Allen: At the end of May the statute of limitations was invoked by the Supreme Court rule and many of the cases are thrown out because they were not instituted within a year after the injury occurred. Of course, the men did not know; it had not been diagnosed; the symptoms develop later in silicosis. The first men who instituted suit and who died have been paid off and some received a part of the first settlement.

[omission]

Mr. Dunn: I understood you to say yesterday that some of the men you approached asked, "What are you going to do for us?"

Miss Allen: Yes; they all say that, everyone.

[omission, page 26]

Mr. Lambertson: We do not know, ourselves, what we are going to do until we hear all of the evidence.

Mr. Dunn: That is correct. We have become interested in these unfortunates because it is not only our duty as persons obtaining a salary for our services, but for the sake of humanity.

Miss Allen: Yes; from the humanitarian viewpoint.

Mr. Randolph: I think all of the committee will commend you for that attitude.

Mr. Dunn: I want to thank you personally for the good work you have done.

Mr. Griswold: Were you ever in the tunnel yourself?

Miss Allen: I went down in 1932 to see what I could and they were just lining the mouth with steel and concrete. I walked down in there among the men while they were hammering the steel together and riveting it.

[omission, page 27]

Mr. Griswold: Have you ever observed the rock structure that they brought out of the tunnel?

Miss Allen: Yes; I have.

Mr. Griswold: What did they do with what they got the rock out of the tunnel from tunneling and blasting; where did they remove it to?

Miss Allen: In tunnel no.1 the silica percentage of the rock was so high it was a very valuable deposit, and they enlarged the tunnel at the point of the most valuable part, where it ran 99.4 percent and loaded it into cars on the C&O tracks, I understand, and took it down to the Electro-Metallurgical plant.

That was in the beginning, when—you see, this was back in 1932, and at that time the New Kanawha Power Co. was, substantially, what I am sure you would call a public utility. It had obtained a license from the State power commission to develop this power for general sale to public firms that wanted it.

Mr. Griswold: They were developing the power. What I am trying to get at, Miss Allen, is did they use this silica that they obtained from the tunnel; did they afterward sell that and use that in commerce? That is what I am endeavoring to discover.

Miss Allen: They used it in the electro-processing of steel.

Mr. Griswold: This particular tonnage that they obtained from this tunnel, did they use that, sell that and use it, do you know?

Miss Allen: It was stored in the yard, and I imagine it is being used in this plant.

[omission, page 27]

Mr. Griswold: Now, as to this disease that they have, will you tell the committee the symptoms of that disease that would be observable by the layman?

Miss Allen: Yes. They are very indefinite, such as heavy breathing, lassitude, red-rimmed eyes, bad eyesight—it affects their eyes. That is all I should say that would be observable.

Mr. Griswold: That is the redness of the rims of their eyes?

Miss Allen: And the general lassitude.

Mr. Griswold: And in its more advanced stages, what symptoms are the most noticeable?

Miss Allen: The falling off of weight, a considerable amount in a very short time, even from the beginning. In the most advanced stage, the men get down so that they have no flesh left on them at all.

As they express it down there, the men get so they are all hide, bone, and leaders, which means he is just skin and tendons and looks like a living skeleton. I took that picture in August of 1934, and I received word that this man died in November, and had lingered on in that condition for about 5 months in bed.

[omission, page 28]

The lungs sound, if you tap on a man's chest, it feels very hard and you hear sort of a metallic sound. Well, it is very peculiar.

Mr. Griswold: What treatment do they use, or have they ever made any attempt to give any studied treatment for silicosis in those places where they do attempt treatment?

Miss Allen: I suppose it is simple a matter of rest where they have not been exposed enough to have scar tissued.

Mr. Griswold: After they acquire it.

Miss Allen: All of the doctors agree there is no arresting the disease. There is no stopping it once it is started.

[omission, page 29-30]

Mr. Griswold: There was a new item, I think in one of yesterday's papers by some member of the contracting firm or the power company in which it was stated that all necessary precautions had been taken that were usually taken in that industry to protect these workmen. Has your foundation, or yourself, made any investigation of the precautions taken in other mines of a like nature?

Miss Allen: The standard method of drilling in a tunnel, as I understand it, is wet drilling to keep down the dust, and they were not following that practice.

Mr. Dunn: That statement was made in court?

Miss Allen: I got it from a lawyer. I could not say whether it was in the court records or not, but I think it is.

[omission, page 30]

Mr. Marcantonio: Now, as a matter of fact, this company made some test drillings before they commenced this work?

Miss Allen: Yes; that is the usual practice.

Mr. Marcantonio: And what was the result of these test drillings with regard to silica?

Miss Allen: Well, they knew, of course, that sinking a test bore down a number of feet to find out what sort of rock they were going to be drilling through and, of course, they discovered this rich silica deposit and charted the course of the tunnel just to get out that silica. I cannot stress the fact enough that it was a very valuable deposit.

Mr. Marcantonio: As a matter of fact they originally intended to dig that tunnel a certain size?

Miss Allen: Yes.

Mr. Marcantonio: And then enlarged the size of the tunnel, due to the fact that they discovered silica, and this company wanted to get this silica out?

Miss Allen: That is true for tunnel no.1.

Mr. Marcantonio: Now, do you know what was done with this silica that was taken out of the tunnel no.1? Was it sold?

Miss Allen: No. It was stored there in the yard at the Electro-Metallurgical Co.

[omission, pages 30-31]

Mr. Marcantonio: Now, aside from the question of dry drilling, which is inherently dangerous, did any of these men ever tell you that they used masks?

Miss Allen: Yes, one colored man, when I visited the colored camp at no.1, which was set back in the hills. A workman up there said he had come at the very end of the drilling of the tunnel and the rumor was going around then that rock dust was killing the men, and he happened to notice engineers, executives, general officials, when they came to the tunnel to investigate, at times some of them were wearing masks, and he said, "I thought it must be a good thing to have, so I went out and bought one, and it cost $2.50 out of my own pocket."

Mr. Marcantonio: But no masks were furnished the men who were actually doing the work?

Miss Allen: Not until the very end of the tunnel work. At the end of that tunnel, I understand masks were given; that is, in the last part of 1932 after suits had been instituted.

Mr. Marcantonio: Do you know whether they were at any time with the exception of the last period which you have just described, ever warned by the company officials that they should go out and buy masks or were masks supplied up until that point?

Miss Allen: Never. All of the men I talked to who instituted suit have made it very clear that they knew nothing about the dangers.

Mr. Marcantonio: But masks were used, however, by executives and engineers who visited that tunnel?

Miss Allen: That is a moot point. As I say, I do know that all of the foremen on the tunnel no.1 job died of silicosis themselves except Charlie Gilmore. I am not sure at this time. I understand he had not died. But the higher laborers of the surveying crew testified they did not warn them. As I say, it is a moot point, but Eddie Clark told me that in this trial—his own words were, "The engineers of the New Kanawha Power Co. used masks when they were gathering samples of rock." He pointed that out during the trial in Fayetteville during 1934.

Mr. Marcantonio: So that, according to him, the engineers did use masks when they went into this tunnel, but no warnings as to the necessity of the use of masks was given nor were masks furnished these men except toward the very last part of the construction of this tunnel; is that correct?

Miss Allen: That is true.
Mr. Marcantonio: That is all.

[pages 32-33 omitted]

MONDAY, JANUARY 20, 1936
STATEMENT OF MRS. CHARLES JONES[25]

Mr. Griswold: The first witness this morning is Mrs. Charles Jones. Mrs. Jones, will you please give your address to the reporter?

Mrs. Jones: I live at Gamoca, W. Va.

Mr. Griswold: Mrs. Jones, the committee has under investigation the silicosis situation at Gauley Bridge, W. Va. Will you tell the committee in your own words anything you know about the situation there?

Mrs. Jones: I lost my three sons, and my husband is in very bad condition as the result of working at the Hawks Nest tunnel. One son was 23 years of age, the other son was 21 years of age, and the other son was 18 years of age. My husband is not able to work. He has not been able to work for quite some time. He has silicosis, according to the doctor. We have been having a very hard time making a living since this trouble came to us. We have been living on $2 a week.[26] There are eight of us in the family.

We have a daughter-in-law with two children, and she lives in a garage. It was a garage, but they have made of it a two-room house now. They, too, get $2 a week. She is Cecil's wife.

Mr. Griswold: Does that $2 a week come from relief?

Mrs. Jones: Yes.

Mr. Griswold: I think probably the committee could proceed better if we used the question method with this witness.

Mr. Lambertson: Mrs. Jones, how long have your sons been gone?

Mrs. Jones: All three of them died within 13 months. Shirley died in 1932, the 18th of June, and on the 25th of September Cecil died. On October 27, 1933, Oren died.

I lost my brother-in-law[27] as a result of his working in this tunnel.

Another boy named Jeffrey,[28] who had made his home with me until 2 weeks before he died, also died as the result of this working in the tunnel.

Mr. Lambertson: How long were they sick, on the average?

Mrs. Jones: Shirley was sick about 3 months, and during that time he could not go

---

26. "An investigation . . . ." pg. 37-41.
27. Two months after testifying before the Congressional subcommittee, the Jones' $2 per week relief check was discontinued.
28. Raymond Johnson.
28. Following the death of his mother, the Jones "took in" Oley Jeffrey. He lived with them for seven years.

around. I would carry him from his bed to the table and from his bed to the porch in my arms.

Mr. Lambertson: How did the silicosis affect them?

Mrs. Jones: They could not breathe. There was a shortage of breath. They complained of a pain in the chest.

Mr. Lambertson: Have conditions gotten better or have they grown worse?

Mrs. Jones: They grow worse all the time. The one boy worked there about 18 months altogether, and he came home one evening with a shortness of breath. He said, "Mother, I cannot get my breath." I told him, "Son, I believe that dust is harming you." I kept him at home after that for a while, and then the tunnel foreman came and asked why the boy was not at work. . . and I told him that I thought the tunnel dust was killing them. He said, "No, that is just a foolish idea of yours. . . ." When the boys would come home they would be all covered with this dust. It would be in their hair, in their eyes, and on their clothes. When they would come home and drop their clothes on the floor the dust would scatter all over the floor from the clothes.

Mr. Lambertson: Did the boys say to you that they thought the law was being violated by the company with dry drilling?

Mrs. Jones: I do not think the boys knew that it was against the law to drill dry.

When they took sick I saw a doctor, but I could not get anything from him. Finally I begged money with which to put them in the Charleston hospital for the purpose of having x-rays made of their lungs. After they were x-rayed, Dr. Harless examined the x-ray pictures. I think another doctor or two examined them and they all said the boys had silicosis. The youngest boy did not get to go down there with me and he said, "Mother, when I die I want you to have them open me up and see if that dust killed me. Try to get compensation, because you will not have any way of making your living when we are gone, and the rest of them are going too." After the death of these two boys we had their lungs removed to determine their condition.

Mr. Lambertson: Where was that done?

Mrs. Jones: It was done in Montgomery, I think.

Mr. Lambertson: Do you know in what hospital it was done?

Mrs. Jones: The undertaker looked after that. Dr. Harless had it done.

Mr. Lambertson: Who examined him? Dr. Harless?

Mrs. Jones: Dr. Harless and another doctor that came in there. I do not remember his name.

Some of them sued, but somehow they must have been bought off. It did not amount to anything. We got $800 on each death. The first one that died asked me to take the money that I received as a result of his death and buy me a home, which I did. I got $800 and $500 and bought a 5-acre farm upon which there was a four-room house. We did not have anything left after that.

I asked the director of relief for help, but they were not willing to give it to me when they learned I had a cow. They asked me how I would keep the cow for $2 a week. I said that one week I would buy feed for the cow and the next week I would buy flour for the children.

Mr. Lambertson: Have you actually sustained yourself on that $2 a week?

Mrs. Jones: I have picked up a washing once a week and made an extra dollar or two that way to help out.

Mr. Dunn: Did I understand you to say that your husband has silicosis?

Mrs. Jones: Yes sir, he has.

Mr. Dunn: Have the doctors given you information as to how long he probably will live?

Mrs. Jones: I asked them about my husband working on relief jobs, and they told me not to have him go to work. They advised me to keep him at home.

Mr. Dunn: In other words, he is too sick to work?

Mrs. Jones: He does not seem to be so sick; he is just short of breath.

Mr. Dunn: How many children have you at home.

Mrs. Jones: We have six children at home

Mr. Dunn: All of whom, I presume, are dependent upon you for support?

Mrs. Jones: Yes; that's right.

Mr. Dunn: The $2 a week that you obtain comes from relief funds?

Mrs. Jones: It does.

Mr. Dunn: Did you ever ask for more? Have you asked for more relief?

Mrs. Jones: Yes, sir; but they thought $2 a week was sufficient. They will not send me any more, or give it to me. I get out on that road and hitch-hike 18 miles at the end of each week to get that relief check. I go with my daughter-in-law who gets $2 a week. She has two children.

Mr. Dunn: You bought a 4-room house?

Mrs. Jones: Yes, sir.

Mr. Dunn: How do you pay taxes on it?

Mrs. Jones: My oldest daughter has janitor work and she paid the taxes from that money.

Mr. Dunn: In other words, the whole family is still in need of more assistance?

Mrs. Jones: Yes, sir; we can hardly live on what we get. I go to bed many nights crying and wondering how I will get food for the next day.

Mr. Dunn: Your three sons died within 2 years?

Mrs. Jones: They died within 13 months.

Mr. Dunn: How old was the youngest son?

Mrs. Jones: He was 18.

Mr. Dunn: How old was the next son?

Mrs. Jones: He was 21 years old.

Mr. Dunn: And the next?

Mrs. Jones: Twenty-three.

Mr. Randolph: Before the sons started working in Hawks Nest tunnel, did you have any illness in your family?

Mrs. Jones: No, sir; we never did. They all had experienced perfect health.

Mr. Marcantonio: It has been stated in some of the newspapers that only three white men died. That statement seems to have been inspired by the contractor who built that tunnel. You say that at least four white men died?

Mrs. Jones: Three of my sons, my brother-in-law, and a boy who made his home with me and whom I kept for about 7 years died. He died with the same thing my boys died with.

Mr. Marcantonio: Do you know how many persons died as a result of working in the tunnel?

Mrs. Jones: I can't say exactly, but there were many of them. Reports kept coming in every day about men working in the tunnel dying. Every day somebody died. Many colored men died.

Mr. Marcantonio: Where did these reports come from?

Mrs. Jones: Gauley Bridge.

Mr. Marcantonio: Go ahead.

Mrs. Jones: They call it pneumonia at first. They didn't agree that it was silicosis until my boy died. They would pronounce it fever and that would be all there would be to it. They would then bury the men and forget it. My youngest son asked that we try to find out whether the dust killed him; we did so, and that's how they learned what the trouble was.

Mr. Marcantonio: How far were you living from the tunnel itself at that time?

Mrs. Jones: I can't say exactly, but I imagine it was about 4 miles.

Mr. Marcantonio: Did you talk to many of the men who came there to work in the tunnel, who worked in and out of the tunnel?

Mrs. Jones: Yes, sir.

Mr. Marcantonio: And they told you that the men were dying, did they?

Mrs. Jones: Yes sir. The foreman told me that this was not dangerous work. I asked one of them about the boy's health, and he said I was too fearful for them. The boys kept getting shorter and shorter breath, and I didn't know what else could be wrong with them. The foreman told me that he had worked in tunnels for 30 years and it didn't hurt him.

Mr. Marcantonio: What was his name?

Mrs. Jones: Oscar Anders. His brother too was a foreman and he is dead.

Mr. Marcantonio: I am referring to the reports you heard about men dying. How did you get these reports?

Mrs. Jones: They came to me in different ways. The men were dying and dead, that was all there was to it. The news was general around that part of the country.

Mr. Marcantonio: You say there were reports of many dying as the result of working in the tunnel?

Mrs. Jones: Yes, sir.

Mr. Marcantonio: That was before your first son died?

Mrs. Jones: About the same time they were carrying so many colored men out.

Mr. Marcantonio: As I understand, you have told us that you have to go 14 or 18 miles to get your relief check. Have you asked the relief authorities to mail the checks to you?

Mrs. Jones: Yes, but they said they didn't have the pennies to mail the checks with. They said that the only way we could get the checks would be to come and get them ourselves.

Mr. Griswold: Did you sign for those checks?

Mrs. Jones: No, sir; they were simply made out and handed to us.

Mr. Griswold: They just handed them out and you did not sign anything?

Mrs. Jones: That's right; they just handed them to us and we took them and went on.

Mr. Randolph: I should like to say, simply as an observation in connection with Mrs. Jones' statement about receiving her relief check, that the reason the relief authorities have not mailed the checks to her each week, which was that they did not have the money to pay postage, if followed generally, is certainly a bad practice.

Mrs. Jones: That's exactly the way the thing has been handled and is still being handled.

Mr. Lambertson: Do you suppose that the relief authorities knew that this woman had to hitchhike 18 miles for this check?

Mrs. Jones: Yes. They knew that. I went there and when I got there the check was not made out. They wanted me to come back for it and I asked them how they expected me to come that far for the check.

Mr. Griswold: The testimony of Miss Allen the other day was that the recipients of this relief had to walk about 16 to 18 miles to get it.

[omission, page 41]

Mr. Dunn: You get relief from this one organization and you walk 18 miles to get it?

Mrs. Jones: That is right.

EXCERPTS FROM THE TESTIMONY OF MR. CHARLES JONES[29]

Mr. Griswold: Is Mr. Charles Jones present?

Mr. Jones: Yes.

Mr. Griswold: Please state your name and residence to the reporter.

Mr. Jones: My name is Charles Jones, and I live at Gamoca.

Mr. Griswold: You understand that the object of this committee is to investigate silicosis and its causes at Gauley Bridge and other place in you country. Will you please proceed to tell us anything that you know about the situation at the Gauley tunnel?

---

29. "An investigation . . . ." pg. 41-48.

Mr. Jones: Yes, sir. I worked in there off and on for 14 months. I carried water and helped to carry steel to the drills.

Mr. Lambertson: You mean you carried water for operating the drills or for drinking purposes?

Mr. Jones: It was for both purposes.

Mr. Lambertson: Have you got silicosis?

Mr. Jones: I have got something that checks my breathing. I have been X-rayed and examined otherwise two or three times. The doctors have told me not to even try to work. I have tried at different places to get jobs, but they will not give me any. I tried to get work with the Kopper Coal Company.

Mr. Lambertson: I take it that you could not work very long, anyway?

Mr. Jones: No, sir; I cannot walk far at a time and I cannot climb a hill because it cuts my wind. I tried to get work at the Kropper [sic] Coal Company but they turned me down.

Mr. Randolph: You have mentioned having sought work at the Kropper [sic] Coal Co. Have you been examined by their physicians?

Mr. Jones: Yes, sir; I have.

Mr. Lambertson: What did they say was wrong with you?

Mr. Jones: They did not tell me. They just laid my card down on the table and said "We are through with you."

Mr. Marcantonio: But you have been examined by a doctor, and he told you that you had silicosis?

Mr. Jones: Yes, sir. That is right.

Mr. Marcantonio: What was that doctor's name?

Mr. Jones: Dr. Harless. Dr. Huey and others. Dr. Huey lives at Charleston.

Mr. Lambertson: Did you ever wear a mask while working at the tunnel?

Mr. Jones: No, I did not.

Mr. Lambertson: Did anybody else, to your knowledge, wear a mask while working at that tunnel?

Mr. Jones: I have seen men have masks, but they would simply have them on their breasts. I have seen two men wear them.

Mr. Marcantonio: Why were they?

Mr. Jones: One was an engineer who came there and marked up the heading and centers to drill by. The other man, too, was an engineer.

Mr. Marcantonio: These two men that you saw wearing masks were engineers?

Mr. Jones: Yes, sir.

Mr. Marcantonio: They were not just ordinary workingmen in the tunnel?

Mr. Jones: No, sir.

Mr. Marcantonio: Did any workman in that tunnel wear a mask?

Mr. Jones: No, sir; not one of them wore a mask.

Mr. Marcantonio: When did you start working in that tunnel?

Mr. Jones: I commenced just before Christmas in 1930. I do not remember the exact date, but I got a pay just before Christmas. I had Christmas money. I worked probably 10 or 12 days before Christmas that year. I worked off and on after that until February 1932. I worked the 11th, 12th and 13th of February 1932.

Mr. Marcantonio: Tell us something about some of the conditions in the tunnel at the time you worked there.

Mr. Jones: Well, it was quite cloudy in there. It was smoky. When the dynamite would be exploded there would be smoke and dust around there for quite a long time. After one worked there a little while he began to get sleepy; but the boss would not allow one to quit work or to sit down.

Mr. Marcantonio: What do you mean by saying, "The boss would not allow one to sit down?" What did he do about it?

Mr. Jones: He would fire a fellow if he saw him sitting down on the job when he should be working. Naturally a man would get sleepy and drowsy in those conditions, and whenever the boss caught a man in that condition and not coming up with his end of the work he fired the man. I knew it was head air, because I had worked in the mines and had experience with it. The dry air from the drills made us sick and sleepy.

Mr. Marcantonio: What kind of drilling were they doing?

Mr. Jones: They had a cut of 52 holes and they used 4 drills. I think they were Ingersoll drills. They had two of them on each post. They drilled holes straight in, in a circle, and they drilled [pointing] this way and then they filled the holes with dynamite and shot them off. They drilled 24 holes straight down [indicating]. They drilled that many or more 12 foot ones, and on the next bench they would level up the bottom for the shovel.

Mr. Marcantonio: Was this dry drilling?

Mr. Jones: Yes. They had four drills with water at the head.

Mr. Marcantonio: How soon after the explosion would take place were the men required to get back to work?

Mr. Jones: The day crew would shoot. They would come out at 6 o'clock. As they came out they would shoot. At 6:30 o'clock we would go in and start work.

Mr. Marcantonio: What was the condition of the air where you were in there that soon after the shooting?

Mr. Jones: It was as full of smoke as it could be. I ran right up upon the shovel and busted my head against it, simply because I could not see it for the smoke and dust.

Mr. Marcantonio: So that unless the headlight was turned directly on the shovel you could not see it?

Mr. Jones: That is right. There was a good light every 10 or 15 feet. These lights would have bulbs of 50 or 100 watts, but right next to the shovel one could not see even with those good lights. Only when they would center the high-powered lights on them could you see.

Mr. Marcantonio: While you were working there did you acquire any knowledge as to men dying?

Mr. Jones: None of the men died while working there, but after they drove a heading through they began dying. About the time the heading was done the men commenced to die.

Mr. Marcantonio: Were you working there at that time?

Mr. Jones: No, I was not working there when my boys died. I had quit at that time.

Mr. Marcantonio: You quit working there after your son died?

Mr. Jones: No, I quit before that. We finished up the heading and then they laid us off. Shirley was the first of our family to die. Several colored fellows died before that. Shirley was first to die, then Cecil died, and then Jeffrey died, and then Oren, and then Raymond Johnson, and then Clev. Anders, Oscar Anders, Frank Dickinson, Frank Lynch, Henry Palf, Mr. Wall, who was assistant superintendent, Mr. Fitch, a foreman. All of these men were white. There was a slim fellow who carried steel with my boys. His name was Darnell, I believe. He, too, died. I could not tell you the number of colored men that died on that job.

Mr. Marcantonio: You would say that there were quite a few colored men that died on that job?

Mr. Jones: Yes, sir.

Mr. Marcantonio: I understood you to say that all the men you mentioned a little while ago were white men?

Mr. Jones: Yes, sir.

Mr. Marcantonio: They are dead?

Mr. Jones: Yes, sir.

Mr. Marcantonio: Were they old men or young men?

Mr. Jones: I do not believe any of them was more that 50 years of age.

Mr. Marcantonio: Were many of them in their thirties?

Mr. Jones: Yes, sir. My youngest son was 18 or 19 when he went to work in the tunnel.

Mr. Marcantonio: And you distinctly remember these two engineers wearing masks, do you?

Mr. Jones: I do.

Mr. Griswold: You have a noticeable wheeze when you breathe. Do you have that all the time or just today?

Mr. Jones: All the time. I do not notice it.

Mr. Griswold: Did you have that before you went to work at the tunnel?

Mr. Jones: No; I did not.

Mr. Griswold: Your breathing was clear prior to that time?

Mr. Jones: Yes.

Mr. Griswold: Did the other members of your family who worked in the tunnel have this wheeze after they worked in the tunnel?

Mr. Jones: Yes, sir.

Mr. Dunn: Was the doctor who examined you and told you you had silicosis, a company doctor?

Mr. Jones: He is in a way and in a way he is not. When we were examined we had to sign a statement, but they did not collect it. We signed a statement that we would pay him $50 each out of our checks if we got them.

Mr. Dunn: In other words, you agreed to pay $50 for one examination?

Mr. Jones: Yes, but he was to take us and have us X-rayed. Dr. Hayhurst examined my boys and Dr. Huey and some other fellow that travels through the country there on a railroad car examining people by X-ray. He took two or three X-ray pictures for him.

Mr. Dunn: Were you taken to a hospital when you were examined?

Mr. Jones: Yes, I was taken to the General Hospital at Charleston. There was an examination made in this railroad car, too.

Mr. Dunn: Do you know whether or not this doctor who examined you was in the employ of the company at the time of the examination?

Mr. Jones: Yes. He was employed by the Midvale Colliery Co.

Mr. Griswold: Do you know what compensation Dr. Hayhurst received for his examinations?

Mr. Jones: No; I do not.

Mr. Griswold: Did Dr. Hayhurst examine quite a few men who had worked in the tunnel?

Mr. Jones: Yes, he helped examine all the X-rays when Rinehart and Dennis paid off.

Mr. Griswold: How were these company doctors paid? Did the men themselves pay them?

Mr. Jones: Yes sir, they did.

Mr. Griswold: How did they pay—was the amount deducted from your wages?

Mr. Jones: Yes.

Mr. Griswold: How much was deducted from your wages for payment of these doctors?

Mr. Jones: A quarter a week and then there was a quarter a week taken for the hospital fund.

Mr. Griswold: There was a quarter a week for treatment by the physician and a quarter a week for a hospital fund?

Mr. Jones: Yes.

Mr. Griswold: Did you ever receive any hospitalization?

Mr. Jones: Yes, for my daughter. She had appendicitis and there was a 12-year-old boy who received help from him.

Mr. Lambertson: How much did your family get for the death of your boys?

Mr. Jones: We received $800 apiece. . . we got it for Shirley, and I got $800 for myself.

[omission, pg. 45]

Mr. Lambertson: You have received $800 each, including yourself, from the company?

Mr. Jones: I got the $800. When I got my check I owed the Midvale Colliery Co. for furniture, rent and coal.

Mr. Lambertson: How much did you owe the Midvale Colliery Company?

Mr. Jones: I owed it $339.

Mr. Lambertson: Has all that money been expended?

Mr. Jones: Yes, it has.

Mr. Lambertson: You received $2,400 on account of the boys and $800 on account of yourself?

Mr. Jones: Owen gave us $500 and the one we collected on ourselves made $1,500. I spent mine. I paid the Midvale Colliery Co. $339. I had to sign a contract with that company if I stayed in the house. If I did not sign that contract I would have to move out.

Mr. Lambertson: When did you get the money for the boys and yourself?

Mr. Jones: I believe it was in July 1933, but I am not sure of that date.

Mr. Griswold: Did you buy the house your wife spoke of with this money?

Mr. Jones: Yes, I did.

Mr. Lambertson: How much did you pay for the house?

Mr. Jones: I bought the house, two cows and a heifer. I paid $1,700 for the place.

Mr. Lambertson: The money is all gone?

Mr. Jones: Yes.

Mr. Lambertson: Is the farm mortgaged?

Mr. Jones: No, sir. We do not owe anything on it.

Mr. Griswold: The $339 plus $1,700 you paid for the place accounts for $2,039 of the money you received, does it not?

Mr. Jones: I did not tell you about my other debts. I had to pay Mrs. Dunbar $1,100, Mr. Olstead $10.50, the Gilbey grocery man $40 and a little more. I had to pay Nancy Miller $15 for rent.

Mr. Griswold: These were the bills that had accumulated after you left the tunnel employment and while this matter was pending against the company?

Mr. Jones: Yes sir.

Mr. Griswold: Was $800 the amount received by you or was that the amount the company paid?

Mr. Jones: I do not understand the questions.

Mr. Griswold: Do you know how much the company paid on account of these claims? Did they pay anything that was deducted before [you] got your money?

Mr. Jones: The attorneys gave us the checks at Lane's [sic] Garage.[30] Each received a check for $800. That was the amount for each death. I paid $1,700 for the farm.

Mr. Lambertson: How big was the farm for which you paid $1,700?

Mr. Jones: The deed calls for 4 acres, more or less.

Mr. Lambertson: Why did it cost so much, about $400 an acre?

Mr. Jones: I do not know why it cost so much. It was sold before we got it for $3,000.

---

30. Layne's Garage.

Mr. Lambertson: How big is the house?

Mr. Jones: It is a four room house.

Mr. Lambertson: Is it what you would call a good house?

Mr. Jones: Yes, sir; it is a good house.

Mr. Lambertson: And you have 4 acres of land?

Mr. Jones: The land is located right on the highway.

Mr. Lambertson: Is it all good fertile land?

Mr. Jones: No. The lower side of it has 11-foot props under it and the upper side is on the ground.

Mr. Lambertson: One could buy an ordinary farmhouse, such as you have described, for $50 or $100 in most sections of the country. I do not see why the land should cost $1,700 for only 4 acres.

Mr. Marcantonio: Is it a fact that your family was one of the first to receive payments in settlement of claims?

Mr. Jones: Yes.

Mr. Marcantonio: Have you any knowledge as to how much some of the other people received, whether they received as little as $800?

Mr. Jones: Yes sir. I have seen some of the checks involved. I cannot, however, [re]call [sic] names. Some received $300, some received $250. Some received only $60.

[omission, p. 46]

Mr. Marcantonio: Have you any idea how much they paid some of the colored men?

Mr. Jones: I have seen some of those checks amounting to $250, and I have seen them arrested because they had not paid house rent. They collected house rent from those checks.

Mr. Marcantonio: House rent was paid out of the check for $250?

Mr. Jones: Yes. They also paid for coal. Squire Miller and Thomas took the money from the men. Thomas is superintendent of the Midvale Colliery company.

Mr. Marcantonio: Who paid the funeral expenses?

Mr. Jones: The men who worked in the mines with me for 18 years paid that expense. At first I worked in the mine and then when they did not run very steadily I went to work for Rinehart and Dennis for this tunnel work. I had received $1.80 for 8 hours working in the mine and after I got laid off at the tunnel I went back to the mines. I went back there at $1.80 a day. For every day I worked they would give me a dollar and hold 80¢ on account of rent, coal, light and so forth.

Mr. Marcantonio: Who paid for Shirley's funeral?

Mr. Jones: Mr. Deets[31] claims he paid it, but he did not.

---

31. Ruby Winebrenner, Gauley Bridge's resident historian, indicated this name should be spelled "Deitz."

Mr. Marcantonio: Who is Mr. Deets?

Mr. Jones: He is general manager of the Midvale Colliery Company.

Mr. Marcantonio: Has that company any connection with the tunnel that was being constructed?

Mr. Jones: No.

Mr. Griswold: You said that Mr. Deets claimed that he paid the funeral expenses of you son, Shirley, but he did not do so. Who did pay the funeral expenses?

Mr. Jones: Mr. Deets did actually pay over the money but the money was received from the men. Mr. Deets collected money from the men. The men banked the money with him and then Mr. Deets paid it out in accordance with an understanding. The men build up [sic] this fund and when somebody dies Mr. Deets actually pays the money. He does not, though, pay the money from his own pocket.

Mr. Griswold: He is agent for the men and yet he takes credit for having paid the funeral bill?

Mr. Jones: Yes.

Mr. Lambertson: When you got the money did you sign anything for the company?

Mr. Jones: No; not for the company, but we did for the lawyers. We signed some kind of papers saying that for a dollar in hand we did so and so. I cannot tell you just what the paper did say.

Mr. Marcantonio: That is what they call a release, is it not?

Mr. Jones: It provided that for a dollar in hand, of which we acknowledged receipt, we released all claims against Rinehart and Dennis on account of injuries received at the tunnel. It was something to that effect.

Mr. Griswold: You mean that in the release you signed for the company, you said that for a consideration of $1 in hand you would release the company from all claims, or something of that sort?

Mr. Jones: Yes.

Mr. Griswold: You do not know whether your claim was originally settled as an original claim for $800 or whether it was paid out of a general account?

Mr. Jones: No, sir; I do not know that. We wanted them to tell us what they would give us before we signed, but they had boosters hauling people around for Bacon and they claimed that Bacon would pay us $7000 each.

Mr. Griswold: Who was the attorney?

Mr. Jones: Hubert was the man who got up the suits and Hubert and Bacon defended them. I believe it should be said that Bacon is the one that defended them.

Mr. Griswold: One man contracted with you to handle the suit and somebody else handled the suits. He employed somebody of ability to handle the suit for him?

Mr. Jones: Yes, I believe that was the way it was handled.

Mr. Marcantonio: Whom did the firm of Hubert and Bacon represent?

Mr. Jones: They represented my boys. A man by the name of Mason, of Charleston, was my lawyer.

Mr. Marcantonio: You would say from your knowledge of the payments and the settlements made that you fared better than others, would you?

Mr. Jones: Yes sir. I believe that is true. We got Townsend, Bock and Moore but we were not satisfied. Some of the colored fellows were told that Bacon did not give us a square deal. We inquired and they sent Mr. Bock up to see about it. It was found that they had withheld $20,000 and never paid it over.[32] Finally they settled. They sent us a check for Raymond Johnson and my two boys and myself for $125. The money on account of Cecil we gave to his wife, Dorothy. My wife had to go down and be appointed administratrix of the estate in order to get Oren's check.

Mr. Marcantonio: What is the story with the $20,000.

Mr. Jones: It seemed that Rinehart and Dennis gave that money to the lawyers to pay these fellows who had not signed releases, and the lawyers kept it, as we found out before the year was up, and we got Townsend to file a suit against Bacon for perjury or something like that and to sue the company. In place of bringing a suit he collected the money and sent it to us.

Mr. Griswold: What is your condition now compared to the time when you quit working in the tunnel? Is it better, worse, or about the same?

Mr. Jones: It is much worse. When I quit them I could go to the table and do such work as driving mules quite satisfactory, but I cannot do that any more.

Mr. Griswold: Are there any further questions? (After a pause.) If not, we thank you for coming before the committee, Mr. Jones.

EXCERPTS FROM THE STATEMENT OF HIRAM SKAGGS[33]

Mr. Griswold: Please give your name and address.

Mr. Skaggs: My name is Hiram Skaggs and I live at Gauley Bridge, W. Va.

Mr. Griswold: Were you ever any time employed at this tunnel at Gauley Bridge, W. Va., Mr. Skaggs?

Mr. Skaggs: Yes, sir.

Mr. Griswold: Go ahead and tell the committee about conditions in that tunnel. What do you know about working conditions there and about silicosis? We should like to know, first, what you did when you commenced working at this tunnel.

Mr. Skaggs: I went there as a drill mechanic. I worked two weeks in the no. 2 heading, and then they laid me off in order to put another man to work. The Ingersoll-Rand Co. took me to no. 1 and I worked there 4 weeks.

Mr. Lambertson: Did you do dry or wet drilling?

Mr. Skaggs: It was, practically speaking, dry drilling.

---

32. Attorneys for the plaintiffs accepted $20,000 from the general manager of Rinehart and Dennis. Upon hearing this information, Judge Eary ordered that $10,000 of this "gift" had to be distributed among the plaintiffs themselves. The Jones' family received $125 from this distribution.

33. "An investigation . . . ." pg. 48-53.

Mr. Lambertson: Why do you say "practically speaking"?

Mr. Skaggs: They used two types of drills: the drifter type and the sinker type. They could not drill with the drifter type without water.

Mr. Lambertson: Are you affected by silicosis at the present time?

Mr. Skaggs: Yes, the doctors say I am.

Mr. Marcantonio: What doctor said that?

Mr. Skaggs: Dr. Rucker of Charleston.

Mr. Lambertson: How long have you worked in that tunnel?

Mr. Skaggs: I worked there six weeks.

Mr. Lambertson: Is that the total of your work there?

Mr. Skaggs: Yes.

Mr. Lambertson: When was that?

Mr. Skaggs: That was in 1931.

Mr. Lambertson: Have you not worked there since then?

Mr. Skaggs: No; I have not.

Mr. Lambertson: And you still have the lingering effect of that disease?

Mr. Skaggs: Yes. When I was there I took sick. The common term used there to describe the disease is "tunnel pneumonia." For about 5 weeks I was under a doctor's care. The doctor for Rinehart and Dennis told me that if I went back in the tunnel it would surely kill me.

Mr. Lambertson: What were the circumstances connected with your leaving the tunnel work?

Mr. Skaggs: I took sick, and when I got better and was able to work again they would not let me go back into the tunnel.

Mr. Lambertson: Did you ever get any compensation?

Mr. Skaggs: No sir; I did not.

Mr. Lambertson: Do you know a good many men who died as a result of silicosis?

Mr. Skaggs: I know a few, at least. I am acquainted with them or was acquainted with them, and I know they died. I was not there long enough to get acquainted with a majority of them, like the other men did. One of the foreman named Pitts is dead.

Mr. Dunn: Do you know what caused his death?

Mr. Skaggs: The doctors said he died as the result of silicosis. When I was working there with the foreman he told me that he knew the tunnel work was killing him. He told me he was taking medicine all the time he worked there.

Mr. Dunn: Did the foreman tell you that?

Mr. Skaggs: Yes.

Mr. Dunn: Do you know how long the foreman had worked in that tunnel?

Mr. Skaggs: No; I do not. Another man, Mr. C. C. Wall, is dead. He worked in there with me.

Mr. Lambertson: What are you doing now?

Mr. Skaggs: I am selling automobiles.

Mr. Randolph: To the best of your knowledge and information, how many men, in your opinion, have died as the result of working in the tunnel?

Mr. Skaggs: That is something I could not say accurately. From reports that one hears and his own observation of the whole thing, from beginning to end, it seems to me that there must have been nearly 1,000 deaths.

Mr. Randolph: That have died?

Mr. Skaggs: Yes, they have died. I could not say that positively, and nobody else could do so.

Mr. Randolph: Are you a native of that region?

Mr. Skaggs: Yes, I am.

Mr. Marcantonio: Your knowledge is based on what you have heard in that region by circulating among workingmen and people who have lived around there, is that correct?

Mr. Skaggs: Yes sir.

[omission, end of 49 – part of page 50]

Mr. Marcantonio: Please describe conditions in the tunnel while drilling was going on.

Mr. Skaggs: They were very bad indeed. I had to go to work in this tunnel on account of my financial condition. They first day I went to work there I did not think I could remain.

Mr. Marcantonio: Why not?

Mr. Skaggs: The dust was so thick that one could not identify anybody he met when the man was only a few feet from him. Then there was a terrific noise, the danger of falling rock, explosions as a result of dynamite that was being touched off. It puts fear into a man and he at once came to the conclusion that he could not stay around there.

Mr. Marcantonio: A statement has been issued to the press, and I have read it, by some official of the contractor in this case, the contractor being Rinehart & Dennis, to the effect that the air down there in that tunnel was as clear as the air on Fifth Avenue in New York City. What have you to say about that?

Mr. Skaggs: That would be impossible, if there was no drilling going on. I have had some other underground experience, and I know that all underground passages contain impure air. I worked in the mines some, but I never worked in one that had a condition as bad as was the condition in this tunnel. The ventilation was not sufficient and the circulation was poor. These drills were running and they created such a dust that the air was completely saturated with the dust. Again, gasoline locomotives were running in there. They had a steam locomotive running up into this tunnel that intensified the difficulty.

[omission, page 50]

Mr. Randolph: Do you think that the contracting firm had competent engineers to handle that situation, or were they given orders not to correct the abuses which, you say, existed?

Mr. Skaggs: That is a matter of opinion, in my judgment.

[omission, page 51]

Mr. Griswold: I notice that you have the same sort of wheeze in your breath that one of the other witnesses has, though it is not quite so pronounced. How long have you had that wheeze?

Mr. Skaggs: Ever since I took sick in the tunnel.

[omission, page 53]

Mr. Griswold: Are there any further questions? (After a pause.) If not, we thank you kindly for coming before the committee.

EXCERPTS FROM THE STATEMENT OF ARTHUR PEYTON[34]

Mr. Griswold: Please give your name and residence to the reporter.
Mr. Peyton: My name is Arthur Peyton, and I live at Glen Ferris, W. Va.
Mr. Griswold: Did you at any time work in the Gauley Bridge Tunnel?
Mr. Peyton: Yes, sir.

[omission, page 53]

Mr. Griswold: Did you work right in the tunnel?
Mr. Peyton: I did.
Mr. Griswold: Go ahead and tell the committee about the conditions you found in the tunnel at the time you worked there.
Mr. Peyton: When I worked there conditions were very bad indeed. The dust was very bad; the circulation of air was very poor; gasoline motors were employed there and they put out an awful fume. The air was very hard to inhale.
Mr. Griswold: When did you work there?
Mr. Peyton: I worked all through the driving of the tunnel from 1930 to 1932.

[omission, page 54]

Mr. Griswold: Are you afflicted by silicosis?
Mr. Peyton: Yes, sir.

[omission, page 54]

Mr. Randolph: Mr. Peyton, Mr. Skaggs who preceded you, has spoken about the failure of the engineering corps of the contractors to provide precautionary measures to

---

34. "An investigation . . . ." pg. 53-66.

make the tunnel working conditions as good as possible; is that your feeling also?

Mr. Peyton: Yes, sir. The contractor for whom I worked, the New Kanawha Power Co. furnished us with respirators. The men who were working with Rinehart and Dennis were not furnished with respirators. The men who work for Rinehart and Dennis were not afforded any precautionary measures to help.

Mr. Randolph: Did you have any conversation with any engineer representing the contracting firm of Rinehart and Dennis about conditions in the tunnel?

Mr. Peyton: Yes, I have talked to the foremen.

Mr. Randolph: Did any foreman admit to you that conditions in the tunnel were bad?

Mr. Peyton: Yes. They just laughed about it though.

Mr. Randolph: They did not take any measures to correct the conditions?

Mr. Peyton: They did not.

Mr. Randolph: Did the Bureau of Mines of the State of West Virginia go into that tunnel and make a report to Rinehart & Dennis that the conditions in the tunnel were not good?

Mr. Peyton: Yes, sir; it did. The evidence was put in court in the case of Raymond Johnson and Donald Shea. Those cases were heard in the Fayette County court.

[omission, pg. 55]

Mr. Randolph: And the contracting firm of Rinehart & Dennis was informed of these bad conditions in the tunnel, you say, throughout?

Mr. Peyton: Yes; that was placed in evidence in the trial of some of these cases.

Mr. Marcantonio: Have you any knowledge of the foremen being forewarned that the mine inspectors were about to visit the tunnel for the purpose of making an inspection?

Mr. Peyton: Yes, sir.

Mr. Marcantonio: What was the practice by the contractor when the mine inspectors were coming to the job?

Mr. Peyton: . . . when the inspectors would come in the one or ones who were watching would inform the foremen and then the process of operation would be changed temporarily while the inspectors were in the mines. For instance, if the men were doing dry drilling at the time the inspectors came, they would stop gradually and go to wet drilling, and also they would get the gasoline motors out of the heading.

[omission, pg 55-56]

Mr. Marcantonio: When you say that the engineering staff of the New Kanawha Power Co. used respirators, how many men used those?

Mr. Peyton: About 20 men.

Mr. Marcantonio: And the men who worked for Rinehart and Dennis, the tunnel contractors, did not use respirators?

Mr. Peyton: They did not.

Mr. Marcantonio: About how many men were involved throughout the period of the tunnel construction? How many worked for Rinehart and Dennis, the tunnel contractors?

Mr. Peyton: I would say 1,000 men in each shift, at all four headings.

Mr. Marcantonio: So that 2,000 men were not using respirators?

Mr. Peyton: They were not.

Mr. Marcantonio: Why did you people use respirators?

Mr. Peyton: Because our company provided them for us for our safety. The company told us to use them.

[omission, page 56]

Mr. Randolph: You have told us that you have silicosis.

Mr. Peyton: Yes, sir.

Mr. Randolph: Therefore, you feel that the respirator was not of much value?

Mr. Peyton: We were there 7 or 8 months before we had the respirators.

[omission, pg 56]

Mr. Marcantonio: The firm of Rinehart & Dennis, the tunnel contractors, kept on with their dry drilling after you folks began to use respirators?

Mr. Peyton: Yes, that is true.

Mr. Marcantonio: And the tunnel contractors, Rinehart & Dennis, kept on using gasoline motors?

Mr. Peyton: Yes.

[omission, pg. 57]

Mr. Marcantonio: How many men do you believe died in the construction of this tunnel as a result of silicosis?

Mr. Peyton: I could not say; but I know quite a few white men died as a result of it.

Mr. Randolph: How many have you actual knowledge of, approximately? The reason I am asking that is you live right in that vicinity?

Mr. Peyton: I do. I have actual knowledge of 15.

[omission, pg. 57]

Mr. Marcantonio: Have you any knowledge as to the number of colored men who died as the result of this tunnel operation?

Mr. Peyton: I would say that between 300 and 1,400 died as a result of that.

Mr. Randolph: White men, do you mean?

Mr. Peyton: I mean white and colored, all told.

[omission, pg. 57]

Mr. Dunn: Can you substantiate the statement that you were unable to see 10 feet in front of you, although there were lights in the tunnel?

Mr. Peyton: Yes. It was very difficult for us to see more that 10 feet when drilling was going on in the tunnel.

Mr. Dunn: It has been testified here by Miss Allen that she was informed that many men working in that tunnel were driven back into the tunnel under most dangerous conditions. Do you also make that statement?

Mr. Peyton: Yes, sir; I do.

Mr. Dunn: Tell us something about it.

Mr. Peyton: Several times after the blasts had been shot off in the tunnel, just a few minutes after that, the foreman would start the colored men back into the tunnel. They would say, "Come on, old Negroes, lets get back to work; let us get out of here." It seemed that the contractors were trying to drive ahead and get through the quickest and cheapest way they could, regardless of human welfare.

Mr. Dunn: Did you say that the foremen employed by the tunnel contractor would drive the colored men right into the tunnel whether they wanted to go or not?

Mr. Peyton: Yes, sir; in the face of the results of those blasts.

[omission, pg 58]

Mr. Dunn: Is it a fact that 10 or 15 of these workers were compelled to sleep in a small shack 10 by 12 feet?

Mr. Peyton: Yes, that is true.

Mr. Griswold: The company charged shack rent for these quarters that were provided for the workers, did it not?

Mr. Peyton: It did.

Mr. Marcantonio: Will you please describe those living quarters.

Mr. Peyton: The shacks in which the workers lived were about 12 by 15. I do not know how many workers lived in each shack, but there were quite a few of them. At the end of each shift, the company would give the colored men, and they may have given some of the white men, or all of them, a check for $3 for working 12 hours. They would give them a check for $3 at the end of each shift, and then charge each worker 10 percent for cashing the check.

Mr. Dunn: Who did that?

Mr. Peyton: Rinehart & Dennis, the tunnel contractors.

[omission, pg 58]

Mr. Marcantonio: Besides charging the workers 10 percent for cashing the checks, they also charged the workers for shack rent?

Mr. Peyton: They charged them shack rent, doctor's fees, hospital fees, and I think they charged some of them for electricity.

[omission, pg 59]

Mr. Marcantonio: The meters governing this electric current were not in the shacks, were they?

Mr. Peyton: They were not.

Mr. Marcantonio: A statement has been made by Rinehart & Dennis, the tunnel contractors, that there is a silicosis racket on, that the best equipment obtainable, the best modern equipment available, and the most efficient equipment available, was used on this tunnel job. You, Mr. Peyton, are an engineer. What is your opinion with reference to the efficiency and excellence of the equipment used on that job?

Mr. Peyton: I think they had very poor equipment indeed.

Mr. Marcantonio: Please give us a detailed statement as to why, in your opinion, that equipment was poor and inadequate.

Mr. Peyton: In the first place, their ventilation system was very poor, indeed. It was too small to take care of the number of men that worked in that tunnel. They had about an 18-inch fan and a vent tube that was about 3 feet in circumference.

[omission, pg 59]

Mr. Peyton: In mine work, they have a large fan. They drive an air current about the same size as the mine, and the air goes all through the place and makes it habitable. There is no such provision in a tunnel operation. There is no favorable comparison between the two systems.

[omission, pg 59]

Mr. Marcantonio: What else have you to say with regard to the equipment in question?

Mr. Peyton: The vent tube used in there all the way through the tunnel was full of holes where rocks had fallen down.

Mr. Marcantonio: Was that not repaired?

Mr. Peyton: It was not; and naturally, this air would come out of these holes in this vent tube and it would not get to the men at the heading.

Mr. Marcantonio: So that you, as an engineer, tell this committee that the air would not reach the place where the men were working?

Mr. Peyton: The air did not reach the place the men were working, because it came out the holes where rocks had torn into the vent tube all the way through.

Mr. Marcantonio: Referring to the matter of drilling, did you observe what kind of drilling was being done?

Mr. Peyton: I did.

Mr. Marcantonio: Was it dry or wet drilling?

Mr. Peyton: The heading drills, drilling straight ahead, were wet. The bench drills were dry.

Mr. Dunn: How many wet drills were in use?

Courtesy West Virginia State Archives.

Mr. Peyton: I think there were six.
Mr. Dunn: How many drills were in use?

[omission, pg 60]

Mr. Peyton: I think there were about 16.
Mr. Dunn: And six of the drills were wet?
Mr. Peyton: Yes.

[omission, pg 60-61]

Mr. Randolph: Referring to the quality of equipment used, was it old or new?
Mr. Peyton: Most of it was old equipment.
Mr. Randolph: Did the Rinehart & Dennis Co. bid against other contractors for this job?
Mr. Peyton: They did.
Mr. Randolph: The contract was let on a competitive bid basis?
Mr. Peyton: It was.
Mr. Randolph: Was there a provision in the bid that the work had to be completed within a certain number of months?

Mr. Peyton: Yes; I think it had to be completed within 4 years. This was the contract specification.

Mr. Randolph: How long did it take the tunnel contractors to complete the job?

Mr. Peyton: It took about a year and a half to drive through the tunnel.

Mr. Marcantonio: As a matter of fact, the job was speeded up as soon as the Federal Power Commission claimed jurisdiction, was it not?

Mr. Peyton: Yes; I believe that is true.

Mr. Marcantonio: Dry drilling is more speedy than wet drilling, I believe you have told us.

[omission, pg 62]

Mr. Randolph: And you found a difference between conditions on this and other jobs with which you were familiar, did you not?

Mr. Peyton: On these other jobs I was not engaged in tunnel work.

Mr. Randolph: I refer to the matter of handling men.

Mr. Peyton: Yes.

Mr. Randolph: Do you mean that the men were badly handled on this particular job?

Mr. Peyton: Yes; they were handled worse than I had ever seen them handled before.

[omission, page 62]

Mr. Dunn: While you were employed on this particular tunnel job, do you remember seeing any employees being carried out after having been overcome by dust?

Mr. Peyton: Yes, I myself was carried out.

Mr. Dunn: How many times were you carried out?

Mr. Peyton: Once I was carried out, but I have seen quite a few such instances. One night I saw 28 men carried out from no. 1 on account of carbon monoxide poisoning.

[omission, pg. 62]

Mr. Dunn: If the contractors had possessed good equipment, would the gasoline have affected the employees?

Mr. Peyton: The company should not have employed gasoline motors in there.

Mr. Marcantonio: What should the company employed there in place of gasoline motors?

Mr. Peyton: It should have employed battery motors. They did use some battery motors, but it seems that they could not get around fast enough and accomplish the work, get the shift finished in good time and in proper order. There were four different headings, and they had a contest going. Each foreman wanted to see whether he could get the most yardage out in a week. They would employ any means whatsoever to get out that yardage, especially in nos. 1 and 4 headings. They were the headings that did that.

Mr. Griswold: How about the subsistence of the workers who lived in these shacks; did they have a company commissary?

Mr. Peyton: Yes sir.

Mr. Griswold: How were payments made for this food from the company commissary?

Mr. Peyton: As I have stated before, most of the payments were made by check.

Mr. Griswold: Did the company make deductions from the pay rolls and give the workers the remainder, or did the company give them the money and let them pay?

Mr. Peyton: They gave each worker a check for $3 at the end of each shift. It was a small card. These men had to buy something at the store and did not have the money, therefore they had to have these checks cashed, and for cashing them the company charged 10 percent.

Mr. Griswold: Where did they cash the checks?

Mr. Peyton: They cashed them at the commissary.

Mr. Griswold: If the men did not have money, would the company give them credit at the commissary?

Mr. Peyton: The company would not do that.

Mr. Griswold: There was no credit?

Mr. Peyton: That is right. If the men did not have money, they could not get anything.

Mr. Griswold: Were the workers required to buy at the commissary?

Mr. Peyton: I do not think they were.

Mr. Griswold: The workers could spend their money where they pleased?

Mr. Peyton: Yes—but most of them had to buy at the commissary. Apparently the company saw to it that the workers were kept in the hole so that they would have to spend their money at the commissary and cash their checks there.

Mr. Griswold: Was this check a regular negotiable one, negotiable any place, or was it payable at the commissary only?

Mr. Peyton: It was payable at the commissary only.

[omission, page 63 bottom]

Mr. Griswold: Did you know a man by the name of McCloud?

Mr. Peyton: I did.

Mr. Griswold: What do you know about him and his manner of handling the workers?

Mr. Peyton: He was a so-called "shack rouster." He would go around to these camps and when these fellows would not show up for work he would chase them out and make them go to work. I have heard it said that he would chase men out to work who were really sick. I suppose they were sick, but he would make them go to work anyway.

Mr. Griswold: Have you ever seen him do that?

Mr. Peyton: Yes; I have.

[omission, pg 64]

Mr. Dunn: Was McCloud brought in from another state?
Mr. Peyton: Yes; as I remember it, his home was in Georgia or some other southern state.
Mr. Dunn: I say he should have died before he left Georgia.
Mr. Peyton: I myself think so.

[end of excerpt. Testimony ends middle of page 66]

EXCERPTS FROM THE STATEMENT OF GEORGE ROBISON[35]

Mr. Griswold: Please give your name and residence to the reporter.
Mr. Robison: George Robison, Vanetta, W. Va.
Mr. Griswold: Did you work in the tunnel at or near Gauley Bridge, W. Va.?
Mr. Robison: Yes, sir.
Mr. Griswold: Tell the committee what you know about the work in that tunnel and the conditions under which it was performed.
Mr. Robison: I went to work there on September 1, 1931, as a driller. I drilled on the bottom bench. They didn't allow any water on the bench drills. The drilling there had to be dry drilling, because otherwise they couldn't drill fast enough. It was at the head that they drilled with water. We put the holes from 1 to 20 feet, straight down. A fellow could drill three holes dry to one wet: that is, it's about three times faster when a fellow drills dry. In shooting at the head they rushed so that they did not even square at the top. The boss was always telling us to hurry, hurry, hurry. When the rocks were in danger of falling at any time the foreman kept telling us that everything was all right and that we should keep right on.
Mr. Griswold: Did they kill any men by falling rock?
Mr. Robison: Yes. Me and my buddy drilled only 4 feet from two other fellows drilling and those two fellows got killed by falling rock. They were only about 4 feet from me and my buddy. The boss himself was up there about the time, and he got tangled up with the equipment on the ground, and he saved himself only by retreating. The two fellows that had been drilling near me and my buddy never knew what hit them. They got crushed beyond recognition.

[lines omitted]

Mr. Griswold: What, if you know, was the name of the boss?
Mr. Robison: We didn't know his name. We just called him and all the rest of the bosses "Cap." That was all I knew. When we were drilling out that bench there would be 8 or 10 drills going. Some nights there would be 8, some nights 9 and some nights 10. That was caused by some men not showing up for work.

---

35. "An investigation . . . ." pg. 66-74.

If we could drill out that night in time they always were loaded, and the powdermen would come in and blow the dust out and load them. We would get the equipment, carry it back, and everybody would get in the clear for shooting. When the blast went off the boss would call out, "Come, let's go back." They loaded them heavy enough to blow the muck down the tunnel and it stayed in there.

When they would bring in water to drink the dust would settle on top of it and one would have to drink that dust too. When drilling, the hole would go straight down and the air would then force the dust back into one's face.

As dark as I am, when I came out of that tunnel in the mornings, if you had been in the tunnel too and had come out at my side, nobody could have told which was the white man.

[sentence omitted]

The groves near the camps had trees that were all colored with this dust. We all tried to clear our clothes somewhat in the parks or groves so as not to have the dust around the shacks. There was so much dust around those groves or parks that it looked like somebody had sprinkled flour around the place. It really looked pretty. Even the rain would not wash away that dust in the parks or groves.

The camps of the colored men were not close to the camps for the white men. If a colored man was sick and really couldn't go to work in the morning, he had to hide out before the shack rouster came around. That fellow had two pistols and blackjack to force the men to go to work. He was a fat man and we called him what we called most of the other white men around there. "Cap."

Mr. Dunn: Was his name McCloud?

Mr. Robison: Yes, I believe it was that; but we just called him "Cap." We couldn't resist him. As I have told you, he had two pistols and a blackjack. If we didn't go with him or go to work he would club us and make us go, and if we resisted him he would shoot us, so there really wasn't anything to do but to do what he told us to do. The only thing a sick man could do to avoid work was to hide out before this man showed up.

I worked on the night shift a while. We had to pay shack rent amounting to 75 cents a week; doctor bill amounted to 25 cents a week; hospital bill was 25 cents a week; the light bill was 25 cents a week.

Mr. Griswold: When you finished did you owe the company or did it owe you?

Mr. Robison: I didn't get very much, I tell you. I always owed the company at the end of the week. Every morning one got a check. They gave a fellow a check, and it was put through at the window. By the time one paid for three meals and got a pint of moonshine, everything was gone.

Mr. Randolph: Did the commissary sell moonshine?

Mr. Robison: No. When we began to cough we thought we had gotten a cold, and we thought it would be well to take some whisky for it. Then, too, we took the moonshine to cut the cold from the lungs. We got the moonshine every day; and I don't believe we could have stayed there without it.

Mr. Dunn: What did you pay for it?

Mr. Robison: A dollar or $1.50 a pint. There were different sizes for pints. We just had to drink something.

When it got so a worker couldn't make it at all, when he got sick and simply couldn't go longer, the sheriff would come around and run him off the place, off the works. I have seen the sheriff and his men run the workers off their places when they were sick and weak, so sick and weak that they could hardly walk. Some of them would have to stand up at the sides of trees to hold themselves up. And the sheriff and his men could plainly see that the men were sick and unable to go; yet they kept making them keep on the move. The workers were from North Carolina, South Carolina, Georgia, Alabama and Florida.

Mr. Griswold: From what state did you come?

Mr. Robison: I am from Georgia, sir. One man didn't know anybody else there. They didn't have any private houses. Everybody on the job had to live on the company's property. One had to keep going on his job, because many hundreds of men were coming in by the trainload and hoboing it. When a man began to let up for any reason, even though he was so sick he simply couldn't keep going, the company would run him off the job.

T.J. Thomas and a Mr. Detch were operating a coal mine near there and some of these men from the tunnel got located there and they are there yet. Many of the men died in the tunnel camps; they died in hospitals, under rocks, and every place else. A man named Finch, who was known to me, died under a rock from silicosis. I can go right now and point to many graves only two blocks from where I live there now.

Mr. Griswold: How many men did you help bury—about how many?

Mr. Robison: I helped to bury about 35, I would say.

Mr. Griswold: Were they all colored men?

Mr. Robison: No, sir; I helped to bury one of the Jones boys. I helped to bury Cecil.

Mr. Dunn: Have you silicosis?

Mr. Robison: Yes, sir. Once I was big as this (indicating) man, but I have fallen off a great deal, as you can see.

Mr. Dunn: How long did you work in the tunnel?

Mr. Robison: Four months. *[omission]* For 7 or 8 months after I quit the tunnel work, during which time I took medicine, I felt as good as I ever did; but after that I commenced to notice a slowing down. When I walked fast or up a hill I could notice that it cut down my breathing.

I talked to the different ones that were suing on account of their condition in 1932 and they told me I had better go to the doctor for [an] examination. I went to Dr. Harless and he said, "Haven't you been working in that tunnel?" I told him, "Yes," and he said "You have that tunnel dust on your lungs." Then I made an arrangement with a lawyer named Mason in Charleston, W. Va. He told me he was going to get me

$25,000 and he would charge me half of it for getting it for me.

Mr. Randolph: Was that James Mason?

Mr. Robison: Yes. The trials were going on at different places around there, then. Mr. Sumerville [sic], too, had some cases. Dr. Harless stated on the witness stand during the trial that any man who had worked in that dust for only 24 hours was afflicted and he could not be cured. He had a list of men there that he said he could show had died from silicosis. He attended all these 35 men I helped to bury. He wasn't the company doctor at the time that I know of.

Mr. Dunn: This lawyer was to get one-half of $25,000 that he agreed to get for you. Did he take your case to court?

Mr. Robison: No. A little after that they got a settlement and they came around in cars with releases and said if we don't sign this we would not get anything. I wanted to get something at once because I wanted to leave right away. If one signed the release he was given a dollar. He gave me the pencil, and I signed, and he took the pencil back, and I didn't get anything at all, sir.

Mr. Dunn: You just had the "pleasure" of signing the release?

Mr. Robison: Yes. The law firm of Townsend, Bock & Moore had a runner there at the same time.

Mr. Griswold: What do you mean by that?

Mr. Robison: There was a man there boosting that law firm. He said that Mr. Townsend was the best man, the best lawyer, in the State of West Virginia.

Mr. Randolph: Is he the attorney for the United Mine Workers of West Virginia?

Mr. Robison: Yes. He told us to turn all our cases over to Mr. Townsend. He said, "All you men turn your cases over to Mr. Townsend." I turned my case over to Mr. Townsend. The man said, "Mr. Townsend will fix you up." When there was a settlement I went down and he came out with a piece of wide, blue-black paper saying, "You filed too late; the statute of limitations has operated against you and turned you down. I can't do anything for you." And I haven't got anything yet.

[omitted pages 70-71]

Mr. Marcantonio: In your opinion, about how many persons have died from working in that tunnel?

Mr. Robison: That I know personally?

Mr. Marcantonio: That you know of.

Mr. Robison: I know about 118 who have died—close to that, I believe. They are colored people.

Mr. Griswold: Colored people only—you have no knowledge of such white persons?

Mr. Robison: I know several white people who have died from that work. I know some foremen who have died. The foremen had to stay in the dust to keep the men there.

[several lines omitted]

Mr. Dunn: Would you make the same statement as the preceding witness, namely, had the company used wet drills instead of dry ones, there would not have been so many deaths?

Mr. Robison: All wet drills and kept the gasoline motors out of there. The gasoline motors were brought in there. On a cloudy, still day the smoke stays in there, and the smoke will do more harm than a pint of potash whisky. The smoke won't go out. That dampness drives the smoke back and down up and down the tunnel all the time.

Mr. Dunn: The statement has been made before this committee that shortly after men died, within an hour after they died they were buried. Is that a fact?

Mr. Robison: I knew a man who died about 4 o'clock in the morning in the camp and at 7 o'clock the same morning his wife took his clothes to the undertaker to dress her dead husband, and when she got there they told her the husband had already been buried.

Mr. Dunn: Who paid the undertaker's bill?

Mr. Robison: Rinehart & Dennis.

Mr. Dunn: Do you know how much it was?

Mr. Robison: They tell me it was $55.

Mr. Randolph: And the officials had knowledge that the man [was] married, yet he was buried about 2 hours after his death?

Mr. Robison: From 4 to 7 which is about 3 hours. When his wife got down there with his clothes the undertaker said he had been buried already. It would have taken an hour to drive up there and back. The wagon could not have gotten back.

Mr. Randolph: Did the undertaker know the man was married, that he had a wife?

Mr. Robison: He went to the house and got the body. He should have known the man was married.

Mr. Griswold: How long have you had the wheeze?

Mr. Robison: I have been having it now ever since the last of 1933. *[omission, page 33]* It is growing worse every day. At times at night I have to get up so that I can catch my breath. If I remained flat on my back I believe I would die.

[omission]

Mr. Marcantonio: Do you know how many of these colored men who came from the South went back there?

Mr. Robison: A good many of them did. When Hubert and Bacon paid off, everybody who got money went back South. I have got letters from people in the South saying that some of the men who worked in that tunnel work had died after they got back South.

Mr. Marcantonio: Relatives of these colored men—who came to that work and then went back south and died—have written back to West Virginia and said the men have died?

Mr. Robison: Yes, sir.

Mr. Marcantonio: There were about 2,000 persons working at this tunnel?

Mr. Robison: Two thousand or more. They had four heads; then they had men on the outside as well as the inside. They worked all the time, night and day except Sundays and Sunday nights.

Mr. Marcantonio: These workers scattered after the work was completed, did they not?

Mr. Robison: They had to scatter.

Mr. Marcantonio: Why?

Mr. Robison: They ran them out of the houses. They even burned down the houses in which the workers lived.

Mr. Marcantonio: Who did that?

Mr. Robison: After Rinehart & Dennis got through with the houses, if they could not sell them—and they did sell some of them—they would burn them. The company men burned them.

Mr. Dunn: They drove the men out after the tunnel was completed?

Mr. Robison: Yes, sir. They put some of the men in jail because they wouldn't vacate the houses of the company.

Mr. Marcantonio: They jailed those who would not readily get off the company's property?

Mr. Robison: Yes, sir; they jailed them for trespassing.

[omission, page 74]

Mr. Marcantonio: How many persons lived in each shack?

Mr. Robison: Different numbers. In some shacks one could find six or eight and in others only two to a shack. Some would have 12 or 14. They had double decks in the shacks. The shacks were made about 12 by 20 or something like that.

Mr. Marcantonio: Were women and children also living in the shacks?

Mr. Robison: Yes.

Mr. Marcantonio: And they charged you for electricity?

Mr. Robison: Fifty cents a half, or twenty-five cents a week.

Mr. Marcantonio: Fifty cents a head?

Mr. Robison: That is for half a month.

Mr. Marcantonio: That 50 cents was paid by only one person living in a shack or was it paid by each person living in a shack? If five men were living in one shack, would each pay?

Mr. Robison: Yes sir; each one paid.

Mr. Marcantonio: In other words, if a shack held 10 persons the company collected 10 quarters a week for electricity?

Mr. Robison: Yes, sir.

Mr. Marcantonio: That was pretty good business.

Mr. Randolph: How many men do you think died as a result of working in that tunnel?

Mr. Robison: I think there were more than 500.

Mr. Randolph: Colored and white?

Mr. Robison: Yes.

Mr. Dunn: There were more colored persons working in the tunnel than white persons?

Mr. Robison: Yes, sir; very few white people could they get to go into that tunnel.

Mr. Dunn: You think there were more colored people because they could not find employment elsewhere?

Mr. Robison: That company knew what it took to do a job like that. They always employ colored people.

Mr. Dunn: Because they work for less wages?

Mr. Robison: Yes sir; and it is faster and quicker labor.

[last paragraph omitted]

LETTER TO WILLIAM CONNERY, CHAIRMAN, FROM DR. L.R. HARLESS, M.D., COMMITTEE ON LABOR[36]

January 19, 1936
Hon. William H. Conners [sic],[37]
House of Representatives Office Bldg.,
Washington, D.C.
Dear Sir:

Due to the illness of my wife and urgent professional duties, I am unable to appear before your Committee on January 20 as per your request by telegram.

I presume your Committee wishes to obtain certain information from me in regard to certain conditions arising from the construction of the Hawks Nest power development near Gauley Bridge, W. Va. Therefore, I am putting down in this letter the information which I would give you verbally were it possible for me to appear before you.

First, permit me to say that this situation at Gauley Bridge, W.Va. has been grossly exaggerated by statements appearing in certain newspapers and by wild rumors current in this section. No doubt these statements have left in the public mind an exaggerated and erroneous impression of the conditions as they actually exist. In order to clarify the situation, here are the facts, as in my opinion and judgment, they really exist.

---

36. "An investigation . . . ." pg. 79-80.
37. William H. Connery.

In June of 1933 a medical commission composed of Dr. Emery R. Hayhurst of Columbus, Ohio; Dr. W.R. Hughey of Charleston, W.Va., and myself were designated to examine a large number of claimants who had filed suit against the Contracting Company for damages resulting from the alleged affection of silicosis contracted in the construction of the tunnel, and these suits were settled out of court. At the conclusion of our examinations we found that out of 337 claimants 13 had died from silicosis and that 139 had some lung damage, by reason of dust, ranging from very moderate to advanced silicosis; many of these cases showed co-existing tuberculosis. Since the examinations made by this Commission there have been two additional deaths which have come under my personal observation, which makes a total of 15 deaths resulting from silicosis.

I have seen statements in the Press to the effect that 476 persons have died from silicosis and that 1500 to 2000 are affected with the disease and doomed to die, all, of course, former workers in the tunnel. This, in my opinion, is a gross exaggeration of the true conditions. I am at a loss to know where and how those figures were obtained.

From the group of 307 claimants, referred to above, I examined a large number of these workmen, perhaps as many as 200, on most of whom I kept no record, who claimed to be affected by reason of their employment. I found very little if any impairment of their health which I could attribute to their work in the tunnel. I advised most of them that they had no grounds upon which to base a claim for damages against their employer. However, some of these men, as I later learned, did file suit for damages.

At this time there are to my knowledge only a few cases of silicosis in this community and these are only moderately affected. To my knowledge the last death resulting from silicosis in their community, occurred on November 30, 1934.

It has been said that none of the men knew of the hazard connected with the work. This is not correct. Shortly after the work began many of these workers came to me complaining of chest conditions and I warned many of them of the dust hazard and advised them that continued work under these conditions would result in serious lung damage. Disregarding this warning many of the men continued at this work and later brought suit against their employer for damages.

While I am sure that many of these suits were based on meritorious grounds, I am also convinced that many others took advantage of this situation and made out of it nothing less than a racket.

In this letter I have endeavored to give your committee the facts which came under my observation in connection with this situation.

If I can supply further information I shall be glad to do so.

Respectfully yours,
L.R. Harless

EXCERPTS FROM THE STATEMENT OF WILLIAM J. FINKE[38]

Mr. Griswold: Will you please state your name, residence and occupation?

Mr. Finke: My name is William Finke. I am a merchant, and I live at the Plymouth Hotel in New York City.

Mr. Marcantonio: Did you have any conversation with Dr. Harless in West Virginia? If so, when?

Mr. Finke: About the first week in November, last, I went to West Virginia at the suggestion of the People's Press to get this big story and to take a fair picture and learn what I could about this situation. I did not at the time think that this would run into such great length. I remained there 10 days and spent considerable time with Dr. Harless. He has three lists of names, in addition to records and jars containing the preserved lungs and x-rays and other records. He showed me one list containing 307 names of men who, he said, had died of silicosis. He showed me another list of about 250 names of men who he had examined and found to have silicosis. In their behalf lawsuits were pending. They had not yet died.

Many of those men were living in and around Gauley Bridge. Some of them had gone back to their homes, but they were still suing Rinehart & Dennis, the tunnel contractors.

Dr. Harless had another list of more than 200 names of men whom he had examined once or twice and found them to possess symptoms of silicosis or other lung disorders. Most of those men had gone away. He told me that he had received letters from them from Johns Hopkins, from doctors in medical centers in New York City and elsewhere. Dr. Harless said he had those letters. I am sorry that I did not see the importance of the story at the time I was there, because I believe I could have gotten those records. At the time Dr. Harless had not been interviewed during more than a year, and he was, so to speak, off guard.

Mr. Dunn: Did Dr. Harless seem willing to give you the information?

Mr. Finke: Yes, he seemed very willing to give me information at the time. I do not understand his present attitude, judging by the statement that has just been made.

Mr. Dunn: His statement about those workers having silicosis does not agree with the statements in his letters. Did you hear his letter read?

Mr. Finke: Yes.

Mr. Dunn: What do you think about it?

Mr. Finke: I think it is ridiculous. I can, though, see why he has written a letter of that kind. He will not come here to this committee without compulsion, I am sure.

[omission, pg. 145]

Mr. Dunn: Perhaps somebody put pressure on the doctor.

---

38. "An investigation . . . ." pg. 114-116.

Mr. Finke: I have no doubt about that.

Mr. Dunn: Perhaps they have tried to intimidate him.

Mr. Finke: Dr. Harless told me that the company knew very well what they were sending those workers into. When I asked him why the men had been taken from the South and brought to West Virginia from Georgia, Alabama, and possibly Tennessee, he told me that the company knew very well what they were putting those men into, because 5 years before that tunneling started he, Dr. Harless, went to New York City to attend a conference of more than 500 engineers and doctors where this subject of silicosis was discussed. Dr. Hayhurst was there and when he was asked to address the conference he chose silicosis for his subject. Dr. Harless himself heard that lecture.

I have since learned that every civil and mining engineer in the world knows what silicosis is. It is in the course they have to pursue. They have to contend with industrial disease and it is only natural for them to know about silicosis particularly. Even dentists know about silicosis, because they study occupational diseases. Dr. Hayhurst spoke an hour and a half at the gathering in New York on the subject of silicosis.

Mr. Griswold: Do you mean that this is the same Dr. Hayhurst who was a member of the board of medical officers who examined the men in West Virginia, and that he spoke to that gathering of doctors and engineers in New York City for an hour and a half on the subject of silicosis?

[omission, pg. 116]

Mr. Finke: Yes; he did. He did not anticipate talking on the subject of silicosis especially. There may have been a reason why he picked the subject. Dr. Harless himself told me that Dr. Hayhurst spoke for an hour and a half to those doctors and engineers, all of who were connected with the Union Carbide & Carbon Co. or its subsidiaries.

[remainder omitted]

EXCERPTS FROM STATEMENT BY RUSH DEW HOLT, A SENATOR OF THE UNITED STATES FOR THE STATE OF WEST VIRGINIA[39]

[omissions at beginning]

Mr. Dunn: Senator Holt, I want to thank you for the interest you have shown and are showing in behalf of those worthy unfortunates in your State. I have no doubt that if the resolution before this subcommittee passes the House that you will do your utmost to see it passes the Senate.

Senator Holt: I certainly shall do everything I can to see that it passes the Senate. This is a tragedy of which I am truly ashamed. I am very sorry that anything like this should have happened in the State of West Virginia. I am sure that the majority of West

---

39. "An investigation . . . ." pg. 121-129.

Virginians feel that way about it. I do hope that citizens of the State of West Virginia will not be held responsible for the actions of these absentee landlords.

[omission, pg. 129]

Senator Holt: The present compensation law of West Virginia provides, in connection with silicosis that the workers must have been exposed to silicosis a certain number of months—17 months, I believe, before they may be compensated. That means in this case that they cannot be compensated. Probably only a few of them worked steadily as long as 17 months.

Mr. Griswold: Do you mean to say that the West Virginia Compensation Act provides that men must be in contact with the conditions that caused silicosis for 17 months before they may receive compensation?

Senator Holt: The article about workmen's compensation in West Virginia which Mr. Marcantonio has just handed me provides that a worker must have been employed for 2 years in the same employment.

Mr. Griswold: The workers must be employed 2 years in contact with this silicate dust before they may file a case before your compensation board and get judgment?

Senator Holt: This article provides that if a worker has worked for 2 years in the same employment, if he presents his claim within 1 year after leaving his job, if he has given his life's history in all detail to the employer, if he has never broken any safety rules, he may recover $500, perhaps $1,000, but no more though he should later be dying of silicosis. In other words, a man may get $500 for sacrificing his life for these people.

Mr. Griswold: Provided he has worked 2 years.

Senator Holt: Yes. This tunnel was started in 1930 and not many men worked in it 2 years.

Mr. Dunn: If one worked there a year and 11 months, he is outside the benefits?

Senator Holt: Yes; he is beyond the reach of this help.

Mr. Dunn: 'Man's inhumanity to man makes countless thousands mourn.'*

[omission, pg. 129]

Senator Holt: Yes. This is the most barbaric example of industrial construction that ever happened in this world. That company well knew what it was going to do to these men. They brought in those transients, especially from the South, and treated them worse than dumb animals should be treated. The company openly said that if they killed off those men there were plenty of other men to be had.

[remaining few paragraphs omitted]

---

*Mr. Dunn is quoting Robert Burns, the 18th century poet and lyricist.

EXCERPTS FROM STATEMENT OF JOHN W. FINCH, U. S. DEPARTMENT OF MINES[40]

Mr. Griswold: We have before us now Mr. Finch, Director of the Bureau of Mines of the United States. Mr. Finch, we shall be glad to have you make a statement if you care to do so, or we will proceed in any other way that suits you.

Mr. Finch: I have a statement that does not relate definitely to the situation in West Virginia which you are investigating, but it covers our knowledge of the disease of silicosis.

Mr. Griswold: The committee is very much interested in obtaining same data on the disease itself.

[omission, pg. 129]

Mr. Finch: The health hazards from dusts produced in the mineral industries exceed in true significance the health hazards from gases. The action of gases is usually acute and well recognized, and accordingly considerable attention is given them with the result that fewer workmen as time goes on are exposed to atmospheres of an unhealthful nature. On the other hand, the response to dust exposure is very slow and several years' exposure are required to produce disablement.

The signs of harm, particularly in the early stages, may pass unrecognized by the workmen and his employer. The advanced stages may be attributed to a general failure of health, especially since dust exposure may predispose the worker to other infections and not specifically connected with his occupation.

[omission, pg. 129]

Roughly, there are two general classes of harmful dusts, those that produce respiratory affections and pulmonary diseases, known generally as pneumonoconiosis, and specifically as silicosis, anthracosis, and asbestosis, and those that produce a general systematic poisoning, as lead, mercury and arsenic.

Of all the dust hazards the one of most importance at present is exposure to dust that contains silica. When air that contains silica dust is breathed, the particles are deposited in the lungs where through chemical and pathological reactions and processes they produce a condition characterized anatomically by generalized fibrotic changes. This constitutes a disease known as silicosis. It can be present in degrees varying from imperceptible and apparently unharmful to physical impairment and total disability. In addition to being conducive to physical impairment, it also creates a predisposition to tuberculosis.

Injury from inhaling silica-containing dust is cumulative with exposure, and there are indications that it can be progressive even though exposure is terminated. It can be arrested but is seldom retrogressive to the point of complete recovery. This is

---

40. "An investigation. . . ." pg. 129-138.

important from the viewpoint of liability for compensation because significant injury is demonstrable at any time by X-ray examination.

Silica is one of the most abundant materials in the earth's crust. *[omission]* Sand and sandstone are almost pure silica and granite and igneous rocks are high in silica. Exposure to silica-containing dust is thus inherent in metal mining, rock tunneling in coal mining, quarrying, excavation and sand blasting.

[omission]

From a compensation viewpoint it is at present the most important occupational disease. Owing to the slow development of the harmful stages of silicosis, some industries that have been in existence from 10 to 20 years are just now having their first real manifestations of the disease.

[omission]

Silicosis, formerly termed "miners consumption" was known to the ancients. In a very general way Hippocrates' *Epidemics* speaks of the metal digger who breathes with difficulty and has a pale and wan complexion. Pliny, the elder, in his *Natural History*, speaks of the use of respirators to avoid dust inhalation. Agricola, in his famous work, published in 1556, gives a description of diseases of the respiratory organs. Loehniss, in a book on mines and mine workers, published in 1690, makes reference to the dust causing lung diseases.

In modern times, the first important systematic inquiries were begun in Cornwall, England, by a Royal commission appointed in 1902. The commission found that many of the miners had lung diseases. A similar commission was appointed in the Transvaal in the same year to report on the subject of miners' phthisis. This commission found that the disease was common among rock drillers, and the conclusion was that breathing dust was the cause.

[omission, pages 132-134]

Mr. Finch: The gentleman I called on to answer the question is Mr. William Yant, expert of the Bureau of Mines on dust studies and he is also superintendent of the Pittsburgh experimental station.

[omission, pg. 134]

Mr. Marcantonio: The difference between dry and wet drilling from the standpoint of protecting the welfare and lives of the workers was well known to all persons engaged in mining or tunnel operations after 1914 or 1916, was it not.

Mr. Finch: Yes, that originated in South Africa.

Mr. Dunn: I really think that information was available and generally known in the time of King Tut.

Mr. Griswold: In the early part of your statement you said, remember, that the

Bureau of Mines had recommended wetting in these dust cases.

Mr. Finch: Yes.

Mr. Griswold: Just what authority has the Bureau of Mines to enforce its recommendations?

Mr. Finch: Its original act does not enable the Bureau of Mines to enforce anything; but we have had in more than 99 percent of the cases involved free access to mines for study of these conditions, and our recommendations have generally been accepted and followed.

Mr. Griswold: But if some particular mine or other operation fully saw fit to disobey your recommendations, there was nothing in law by which you could enforce those recommendations?

Mr. Finch: No. It is purely an educational process. We try to give the matter all the publicity we can.

Mr. Griswold: No matter how essential the Bureau of Mines might find certain operation in mining or how essential it might find certain things to protect the health, there is no way the bureau could enforce its recommendation along that line?

Mr. Finch: No; but we can induce them to do it.

Mr. Griswold: In what manner can you induce them to do it?

Mr. Finch: By making them ashamed of themselves.

Mr. Griswold: Some people do not get ashamed of themselves.

Mr. Marcantonio: For instance, Rinehart & Dennis Co.?

[omission, pages 135-138]

MONDAY, JANUARY 27, 1936
EXCERPTS FROM STATEMENT OF JAMES M. MASON[41]

Mr. Randolph: The subcommittee of the Committee of Labor, hearing the witnesses in connection with silicosis trouble in West Virginia, will resume hearings at this time and call to the stand Mr. James M. Mason, an attorney. Mr. Mason, your name has been mentioned in this record as a lawyer practicing at the time of the silicosis flare-up, we will say, silicosis trouble in West Virginia, at Charleston. Your home now is at Charles Town, W. Va., is it not?

Mr. Mason: Yes.

Mr. Randolph: That is about 70 miles from Washington.

Mr. Mason: Yes.

Mr. Randolph: Mr. Mason, tell us in your own way about the silicosis situation there and the victims with whom you came in contact at Charleston at the time, and then we will question you later.

Mr. Mason: In about 1931 I was employed by a committee of the House of

---

41. "An investigation . . . ." pg. 139-147.

Delegates of the State of West Virginia to make an investigation of the workman's compensation department of West Virginia. Following that my attention was called to the working conditions in the Hawks Nest tunnel at or near Gauley, W. Va. That tunnel was constructed in accordance with a contract entered into between Rinehart & Dennis and the New Kanawha Power Co., the latter being a corporation organized for the purpose of constructing that project. That material handled in the construction of that tunnel was 99.4 percent silica.

The working conditions of the men were investigated by the Bureau of Mines, and, if my recollection serves me correctly, at a meeting of the Coal Institute held at Madison, W. Va., Robert Lambie, then head of the Bureau of Mines of West Virginia, made severe criticism and indictment of officials of the Rinehart & Dennis Co., the tunnel contractors. Shortly thereafter that situation had developed, and I was employed by several of the employees of Rinehart & Dennis to institute proper action against Rinehart & Dennis for recovery on account of injuries. Those suits were instituted, and I represented about 96 of the claimants. There were approximately 400 suits instituted then or subsequent thereto.

One trial was had. That trial lasted 10 weeks as I remember.

The information that we got from several of the jurors was that 10 of them had agreed to give the plaintiff in that first case $25,000. Two of the jurors refused to give anything; they refused to consider or discuss the case. Therefore the jury was, of course, discharged from further study of the case. One of those two jurors whose names I do not recall was fined for contempt of court by Judge Erie, because of a reported familiarity or for being allegedly familiar with some of the defendant's employees.

The second case was tried, and at that time 11 of the jurors, according to the information received, voted $25,000, and one juror refused to give anything.

In May or June 1933 I was called to Fayetteville, where these cases were tried, and told that a settlement had been reached. I asked the terms of the settlement and was advised that Rinehart & Dennis agreed to pay $130,000 in settlement of all the cases. I demurred, thinking the amount was not sufficient, and I asked why the sum of $130,000 had been reached. I was told that such was the best figure Rinehart & Dennis had offered or would offer; it was the largest sum that could be obtained.

I refused to settle on behalf of my clients, and I was told that unless I did all papers, all evidence, would be destroyed, and that, of course, meant a helpless situation so far as the clients I represented were concerned. I finally agreed to settle and releases were obtained and the cases were closed in August 1933.

In October 1933 I was advised that Rinehart and Dennis, the tunnel contractors, had paid some of the attorneys for the plaintiff $20,000 in order to effect that settlement. I later found that there was a contract entered into between Mr. Perkins, then general manager of the Rinehart & Dennis Co., tunnel contractors, and two of the attorneys representing the plaintiffs under the terms of which contract Rinehart & Dennis agreed to pay the attorneys $20,000 for effecting the settlement. In May 1934 that contract was a basis upon which to proceed, and I wrote to Mr. Falconer, president of the Rinehart & Dennis Co., tunnel contractors, Charlottesville, Va., and demanded an accounting of that money. I sent a copy of the letter to those representing the defendants and the attorneys who had received the $20,000. I received a letter from Falconer stating that the money had been paid in good faith. One of the two attorneys came to my office and stated that the money had been paid, and that he was in a position where he could raise half of the money that had been paid over, and if we would accept that, the money would be paid. In a few days he brought to us a check for $10,000, and that money was distributed to the clients who had participated in the settlement.

I can only say in passing that I think the payment of that money, the suspicions tampering with the jury system, was about the most damnable outrage that had been perpetrated in any State up to that time.

I think that covers my recital, Mr. Randolph.

Mr. Randolph: Have you, Mr. Marcantonio, any questions?

Mr. Marcantonio: As a matter of fact, Mr. Mason, at one of these trials where the jury was deadlocked, was not one of the jurors held in contempt of court?

Mr. Mason: That is correct.

Mr. Marcantonio: What was the basis of the contempt proceedings against that juror?

Mr. Mason: It was shown to the court that there were many occasions when this particular juror had ridden back and forth from his home to Fayetteville with representatives of the Rinehart & Dennis Co.

Mr. Marcantonio: That was during the trial?

Mr. Mason: Yes.

Mr. Marcantonio: Do you know whether or not an agreement had been entered into where Rinehart & Dennis Co. agreed to pay the attorneys for the plaintiffs the sum of $20,000?

Mr. Mason: Mr. Marcantonio, it was not an agreement by Rinehart & Dennis to pay the $20,000. *[omission, pg. 142]* It was by Mr. Perkins, general manager of Rinehart & Dennis Co., and the other gentleman, whose name I do not recall, who was superintendent for them; but so far as Rinehart & Dennis Co. were concerned the contract did not refer to them as the party paying the money.

Mr. Marcantonio: The contract was one by and between one of the officers of the Rinehart & Dennis Co. and somebody else?

Mr. Mason: Yes.

Mr. Marcantonio: It was between the superintendent of Rinehart & Dennis Co. and the attorneys for the plaintiff?

Mr. Mason: Yes.

Mr. Marcantonio: Have you at any time seen that contract?

Mr. Mason: Yes; I have.

Mr. Marcantonio: State from your memory, to the best of your recollection, the contents of that contract, the terms of the contract.

Mr. Mason: There is a recital of the pending settlement of the cases, and, in consideration of the efforts of these attorneys to bring about the settlement, it provided that they should be paid $20,000 upon consummation of the contract. That is the essence of that contract.

Mr. Marcantonio: These attorneys received their share of the retainer; pursuant to the agreement they received a share of the compromise.

Mr. Mason: They received $30,000 for their fee in that case.

Mr. Marcantonio: And in addition to that they received $20,000 from Mr. Perkins, who represented the Rinehart & Dennis Co.

Mr. Mason: Yes.

Mr. Marcantonio: I should like to make an observation at this time, with the permission of my colleagues.

Mr. Randolph: Proceed.

Mr. Marcantonio: A great deal has been said in the press by or in behalf of Rinehart & Dennis, the tunnel contractors, and particularly by Mr. Falconer, president of the Rinehart & Dennis Co., to the effect that these victims were practicing a racket, that they were practically engaged in a racket, and he, further, described the silicosis situation there as a silicosis racket. I agree with them that a racket has been practiced, but the most damnable racketeering that I have ever known is the paying of a fee to the very attorneys who represented these victims. That is the most outrageous racketeering that has ever come within my knowledge. Not only Al Capone should be in Alcatraz, but these representatives of the Rinehart & Dennis Co., tunnel contractors, should be there with him. Moreover, in order to remove any immunity that may attach to me in this connection, I am going to repeat that on every public platform upon which I appear within the next month or so, I am going to waive all immunity and make that statement and more statements.

[several lines omitted, pg. 142]

Mr. Dunn: The very fact that the corporation decided to make settlement out of court convinces one that the corporation was guilty of violation of law.

Mr. Mason: Yes. I do not think there was any question about the liability of the defendant in those cases.

Mr. Marcantonio: You communicated with Rinehart & Dennis. With reference to the $20,000 it paid to the attorneys for the plaintiff, as I understand you.

Mr. Mason: Yes; I communicated with Mr. Falconer, president of the Rinehart & Dennis Co.

Mr. Marcantonio: Did he answer you?

Mr. Mason: He did. *[omission, pg. 142]* After acknowledging receipt of the letter he said this money had been paid in good faith.

[several lines omitted, pg. 142]

Mr. Marcantonio: He did not deny the payment?

Mr. Mason: He did not.

[several lines omitted, pg. 142]

Mr. Randolph: You mentioned the name of Mr. Lambie, who was at the time head of the State bureau of mines.

Mr. Mason: Yes; he was head of the bureau of mines.

Mr. Randolph: There has been testimony at our hearings that Mr. Lambie changed his position on the question during the course of the trial. Have you any knowledge of that?

Mr. Mason: Nothing except what I learned from his speech at the Coal Institute held at Madison, W. Va., prior to his being a witness in the trial of these cases for Rinehart & Dennis Co.

Mr. Randolph: Can you tell us about that?

Mr. Mason: It has been 3 years since I was connected with those cases, but it has been reported to us, and I think it was carried in the Coal Valley News, that Mr. Lambie made a speech before the Coal Institute at Madison, W. Va., and described the deplorable conditions that these men were required to work under in that tunnel. Later, in his testimony before the court, he described those conditions as being excellent. That is a change of pace, you might say.

[paragraph omitted, pg. 143]

Mr. Randolph: What attorneys received $25,000? What is the name of the law firm?

Mr. Mason: I would rather not give that, Mr. Randolph.

Mr. Dunn: Is there any way whereby we may obtain that information without embarrassing you?

Mr. Mason: Yes; there is.

Mr. Dunn: How may we so get it?

Mr. Mason: It is available at Fayetteville

Mr. Dunn: Is it a matter of court record?

Mr. Mason: No; it is not in the court record.

Mr. Randolph: Was it a Fayetteville firm?

Mr. Mason: Yes.

Mr. Randolph: A firm practicing in the county seat of Fayette County?

Mr. Mason: That is right.

[omission, pg. 144]

Mr. Marcantonio: When did you communicate with Rinehart & Dennis Co. in connection with the payment of $20,000?

Mr. Mason: I do not remember the date.

Mr. Marcantonio: But when you did communicate with Rinehart & Dennis

Co., tunnel contractors, concerning this payment of $20,000 you did receive a communication in black and white from the company stating that the money had been paid in good faith?

Mr. Mason: Yes.

Mr. Marcantonio: That was the only justification that company offered for the payment of the $20,000?

Mr. Mason: The attorneys who were later employed to represent some of the cases were not included in the settlement, and when I gave them the information about that story they confronted Mr. Falconer with it at Charlottesville, which is the main office of the Rinehart & Dennis Co. Mr. Falconer told them the same I had been told—namely, that the money had been paid in good faith—and that was as much as could be gotten from Rinehart & Dennis Co.

Mr. Marcantonio: I want to offer now for the record a letter received by one of the victims of silicosis from the firm of Rinehart & Dennis Co. It reads as follows:

> *Rinehart and Dennis Co.,*
> *Charlottesville, VA., August 26, 1935*
> *Mr. Arthur Peyton,*
> *Dear Sir: Replying to your letter of the 22d, will say that I do not think it would be proper for me to give you the information you asked for, as the transaction we had with Townsend, Bock & Moore was confidential.*
> *Regretting that I am unable to assist you in this matter, I am,*
> *Very truly yours,*
> *E.J. Perkins.*

Mr. Marcantonio: Mr. Peyton had written to Rinehart & Dennis requesting information in connection with some additional money he had received after his settlement.

After Mr. Peyton had received a check marked, "In full payment and in full settlement of all claims afainst Rinehart & Dennis et al.," he subsequently, on June 20, 1934, received the following letter from his attorneys:

> *Lilly & Lilly*
> *Charleston, W. Va., June 20, 1934*
> *Mr. Arthur Peyton,*
> *Glen Ferrie, [sic], W. Va.*
> *(In re: Peyton V. Rinehart & Dennis et al.)*
> *Dear Sir: We take pleasure in enclosing herewith our check payable to you for $21.59, being one-half of the residue which we were able to collect in your behalf in regard to the above case.*

In winding up the various suits, after collecting all we could, we find this balance due you.

With regards, we are
Very truly,
A.A. Lilly

[remaining comments omitted, pg. 145]

LETTER OF REBUTTAL FROM RINEHART & DENNIS CO.

Rather than attending the Congressional hearing, officials of the Rinehart and Dennis Company chose instead to communicate via the following correspondence, dated February 3, 1936. Clearly this company rejected all responsibility for the Hawks Nest disaster, and interestingly also denied any knowledge of the existence of silicosis, although their attorneys previously paid $160,000 in fees to achieve an out-of-court settlement for the silicosis victims.

February 3, 1936 [42]
Honorable William P. Connery, Jr.,
Chairman, Committee on Labor,
House of Representatives,
Washington, D.C.
Dear Sir:

We thank you for the invitation extended to Messrs. Faulconer and Perkins, officials of this Company, to appear before your Sub-Committee on February 5th.

Newspaper reports of your Sub-Committee's hearings show that some of the witnesses, evidently repeating slanderous rumors and hearsay, have made sweeping misstatements which are on their face preposterous and absurd. Some others have made statements which we know to be false, the facts with regard to which can not be adequately treated in the hearings now being conducted. We have, therefore, decided, instead of accepting your invitation, to respectfully suggest that the best way for the Committee to inform itself reliably will be for it to go and see the work we have done and the conditions surrounding it, and at the same time acquaint itself at first hand with the entire situation. We shall be glad to facilitate such a visit.

In the meantime we submit for the Committee's present information the following summary of facts:

Having had considerable experience as tunnel builders, having built fifty-one tunnels within the past thirty years, located in the States of New York, Pennsylvania,

---

42. West Virginia Cultural Center. Union Carbide collection.

Kentucky, Maryland, North Carolina, Virginia and West Virginia, we were one of thirty-five who, in September, 1929 bid for the construction of the Hawks Nest development. After the bids had been compared and our experience, qualifications and character had been investigated, we were, on March 13, 1930, awarded the contract for this work which comprised a dam, a tunnel slightly over three miles long, a power house and some other related features.

Tunneling began June 13, 1930. In May, 1931, at the owner's request, and during the excavation of the tunnel, it was arranged to extend the time of completion to December, 1932, by delaying the work. Again in 1932 the time of completion was extended to January 1, 1934, by further delaying the work. The down-stream section of the tunnel was holed through on August 7, 1931, the up-stream section on September 19, 1931, and the major trimming was finished by December 1, 1931, thereby completing the excavation, which required seventeen and one-half months. However, it was not until December, 1934, that the entire work was virtually completed.

The number of men employed at various times on the whole development was 4,931, of whom 1,687 were white, and 3,244 were colored. The largest number employed at any one time was 1,203, of whom 463 were white and 740 were colored. A total of about 2,500 worked in the tunnel, 500 white and 2000 colored. The largest number on the pay-roll at any one time as tunnel workers was about 600, of whom about one-fourth were white men.

The laborers were partly local residents and partly those who came from elsewhere, seeking work. Some of the latter secured places to live in near-by settlements. Camp houses were built for the remainder. At all times more men were seeking employment that we were able to employ and no labor was ever imported.

Camp sanitation was carefully supervised. An arrangement was made with a first-class hospital for the care of the sick and injured employees, and two doctors were employed, one of whom was resident in the camps. There was a first-aid station, with equipment, at each camp.

After consulting the leading supply and equipment dealers and ventilation experts, we purchased the most up-to-date and improved machinery and equipment on the market. This consisted, in part, of four electric shovels for handling the spoil, storage battery locomotives for use at the headings, gasoline locomotives for hauling out trains of spoil, Ingersoll-Rand air compressors and drills. The drills were all fitted with standard water heads for wet drilling and water pipes and hoses were brought and used to carry water to the drills. Wet drilling was insisted upon at all times. The water supply in ample quantities, under pressure, was available from elevated tanks, which were kept full by constant pumping. Our reason for getting the water equipment and using water on the drills was to make working conditions better in the tunnel, and also because the drilling could be done more efficiently.

The ventilation was taken care of by electric fans and twenty-four inch tubing *which delivered 7,000 cubic feet of air per minute to each heading. The ventilation experts recommended twenty inch tubing, but we decided to use twenty-four inch, which provided fifty per cent more air at the heading.*

Tunneling was done by two shifts daily, except Sunday, when no work was done. A shift worked ten hours or less. The pay was for the full ten hours, even if the shift was shorter, which usually was the case. There was an interval of at least two hours between the cessation of work by one shift and the beginning by the next shift. During this time, both the ventilation tubes and compressed air lines, which furnished air for the drills, were continuously delivering fresh air into the tunnel. Over 8,000 cubic feet of air per minute was delivered at each heading during working hours and between shifts, which was equivalent to more than 150 cubic feet of air per minute per man. Wet drilling was regularly insisted upon and practiced.

Blasting was done at the completion of each shift and at least two hours before the beginning of the next shift. Special gelatin dynamite, designed for tunneling, was employed, and its use was regularly inspected by the manufacturers' experts. Approximately two million pounds of explosives were used without a single fatality. On several occasions, the air in the tunnel was analyzed by chemists. No harmful ingredients from blasting were found, and the average carbon monoxide content was found to be smaller than has been found in busy streets of New York City.

The muck pile of rock broken down by the blasting was moist from the water from the drills and from the condensation of moisture at the headings. This rock was loaded by the electric shovels into cars which were handled by storage battery locomotives, and made up into trains on a switch at a convenient distance from the heading from which the trains were moved to the outside by the gasoline locomotives. This is a standard practice, which has been followed in many tunnels, some of those under Government supervision.

The atmosphere in the tunnel was clear and visibility was so good that long surveying sights were made daily at any time, there being at times some fog from condensation of moisture brought in by the ventilating air.

The accident record on this work compares favorably with that of other undertakings of like size and character. There were seventeen fatal accidents, six of which were on the outside and eleven of which were in the tunnel. Four of those were white and thirteen were colored.

There are official records of eighty-one deaths in West Virginia of persons in our employ, or who had been employed by us, and we have heard of five others outside of that State. Mortality tables in general use by Insurance Companies indicate that out of the 4,931 adult males employed on this job, nearly 300 would be expected to die of natural causes from the beginning of the work up to this time. The recorded vital statistics show that during the years the tunnel was drilled the death rate among our white employees was less than the death rate among the adult white males of

Fayette County at large, and that the death rate among our colored employees was approximately the same as the death rate among the adult colored males of said County.

The methods used in this construction were the standard, or better, that have been used not only by us but by other tunnel builders, both on private and in government projects. We used every safeguard of life and health that was known to us or other contractors in similar work. Conditions were better in this tunnel, and were so considered by many visiting engineers and contractors, than in any other tunnel we had ever seen. We did not furnish nor use dust masks or respirators because no need for them was apparent.

The disease silicosis was not known to us or other contractors of our acquaintance before we were surprised by the bringing of damage suits. We have since learned that there were not available, when the tunnel was under construction, any respirator or mask or other device that would protect against microscopic dust which alone is dangerous and which if present was invisible and unsuspected.

We know of no case of silicosis contracted on this job.

In conclusion we wish to say that our operations were always open to visits and still are. We wish to state that each and every allegation which refers to us and our work contained in the preamble of H.R. Res. 449 is false and the allegation concerning the burial in a cornfield of 169 workers is not only false but absurd. The number buried and the places of burial in West Virginia are definitely and demonstrably known to be otherwise than so described.

Yours respectfully,
RINEHART & DENNIS COMPANY, INC.
By E.H. Perkins, Vice-President
By P.H. Faulconer, President [43]

---

43. P. H. Faulconer and Hollis Rinehart partnered on many financial ventures. Rinehart, "whose family became wealthy from exploits in the railroad business as a partner in Rinehart and Dennis," collaborated with Faulconer on a variety of ventures including underwriting the finances of the Farmington Hunt Club of Greenwood, Virginia and building the Paramount Theatre in Charlottesville—a grandiose theatre that opened its doors on November 25, 1931 during the height of the Depression. One month prior to completion of the tunnel's excavation, while men had already starting succumbing to silicosis, the Paramount opened to rave reviews, described as "the last grand movie palace of the golden age of cinema."

While Rinehart and Dennis Co. purportedly disbanded after the "incident" at Hawks Nest, Faulconer Construction Company, whose equipment can be seen in some of the construction photos, thrives to this day. Whether or not it absorbed the assets of Rinehart & Dennis is unclear; but today Faulconer Construction advertises that it "...has a long and storied history of successfully serving its customers both in Virginia and within the central Atlantic region. With stockholder ties reaching back to the turn-of-the century, when its predecessor company built projects throughout the country...."

SUBCOMMITTEE FINDINGS AND CONCLUSIONS

The following letter-of-finding was issued by the subcommittee upon conclusion of their investigation.

February 5, 1936 [44]
Hon. William P. Connery, Chairman
Committee on Labor,
Washington, D. C.

Dear Mr. Chairman:
The subcommittee appointed to consider H.J. Res. 449, a resolution to authorize the Secretary of Labor to appoint a board of inquiry to ascertain the facts relating to health conditions of workers employed in the construction and maintenance of public utilities, respectfully submits the following report of its investigation:

Your committee held hearings from January 16, 1936, to February 4, 1936, inclusive, and heard many witnesses who testified to the conditions under which workmen were employed at the Hawks Nest tunnel, Gauley Bridge, W. Va.

From the testimony of numerous witnesses, ranging from actual workers on the project to experts from the Federal Bureau of Mines, the subcommittee finds as follows:

That the Hawks Nest tunnel was constructed by the contracting firm of Rinehart & Dennis of Charlottesville, Va., for the New Kanawha Power Co., a subsidiary of the Union Carbide & Carbon Co. That a tunnel was drilled for an approximate distance of 3.75 miles to divert water from New River to a hydroelectric plant at Gauley Junction.

That in most of the tunnel the rock which was drilled contained more than 90 percent silica. That in some of the headings it ran as high as 99 percent pure silica. That this is a fact that was known or by the exercise of ordinary and reasonable care should have been known to the New Kanawha Power Co. and the firm of Rinehart & Dennis.

That silica is a dangerous element to health. That when submitted to contact with silica dust the lungs of human beings become infected with a respiratory disease known as silicosis. This disease is caused by breathing into the lungs silica dust.

That the effect of breathing silica dust is well known to the medical profession and to all properly qualified mining engineers. The disease is incurable. It always results in physical incapacity and in a majority of cases is fatal. That for more than 20 years the United States Bureau of Mines has been issuing warnings and information while conducting an educational campaign on the dangers of silicosis and means of prevention.

That the principal means of prevention are wet drilling, adequate and proper

---

44. "An investigation…." pg. 201-202.

ventilation and circulation or air, the use of respirators by the workmen, and drills equipped with a suction or vacuum-cup appliance.

The subcommittee finds that there was an utter disregard for all and any of these approved methods of prevention in the construction of this tunnel. That the dust was allowed to collect in such quantities and became so dense that visibility of workmen was lowered to a few feet. That workmen left the tunnel at the close of a working shift with their clothing and bodies covered with a dense coating of white silica dust. That the air circulating system was inadequate, insufficient, and out of repair. That respirators were not furnished, or used by, the employees of Rinehart & Dennis. That the majority of drills in use were used for dry drilling. That dry drilling is more rapid and effects a large saving in time and labor cost. That no appliances were used on the drills to prevent concentration of dust in the tunnel. That gasoline locomotives were used in the headings as well as the tunnel entrance and that as a result there was great suffering from monoxide gas among the workers.

That the whole driving of the tunnel was begun, continued, and completed with grave and inhuman disregard of all consideration for the health, lives, and future of the employees.

That as a result many workmen became infected with silicosis; that many died of the disease and many not yet dead are doomed to die from the ravages of the disease as a result of their employment and the negligence of the employing contractor. That such negligence was either willful or the result of inexcusable and indefensible ignorance there can be no doubt on the face of the evidence presented to the committee.

Your subcommittee further finds that the disease of silicosis is prevalent in many States where mine and tunnel operations are now, or have been in the past, in progress. The subcommittee is of the opinion that the investigation thus far has but laid the groundwork and opened the subject for further investigation. That silicosis is one of the greatest menaces among occupational diseases and that State laws governing prevention and compensation are totally inadequate.

It is impossible in this report to go into details concerning all of the testimony. We suggest that the hearings be read in the entirety. The record presents a story of a condition that is hardly conceivable in a democratic government in the present century. It would be more representative of the middle ages. It is the story of a tragedy worthy of the pen of a Victor Hugo—the story of men in the darkest days of the depression, with work hard to secure, driven by despair and the stark fear of hunger to work for a mere existence wage under almost intolerable conditions.

The officials of the contracting firm, Mr. P.H. Faulconer, the president, and Mr. E.J. Perkins, the vice president, were requested to appear before the subcommittee but declined to do so, stating that they had no knowledge of any deaths from silicosis contracted on the work. The record, however, shows that the firm paid some claims for death from the disease.

The subcommittee, therefore, is of the opinion that these officials should be brought before a committee, bringing with them their books and records.

The subcommittee, therefore, recommends that a resolution be presented to the House asking for sufficient funds and authority to require the attendance of witnesses and to do all things necessary to procure a full and complete investigation.

Your subcommittee can do no more. Congress should do no less than to see that these citizens from many States who have paid the price for the electricity to be developed from the tunnel are vindicated.

If by their suffering and death they will have made life safer in future for the men who go beneath the earth to work, if they will have been able to establish a new and greater regard for human life in industry, their suffering may not have been in vain.

Respectfully,
Glenn Griswold, Chairman subcommittee
Vito Marcantonio
W.P. Lambertson
Matthew A. Dunn

LETTER OF OPINION, JENNINGS RANDOLPH

Congressman Jennings Randolph of West Virginia, while not endorsing the subcommittee's findings, instead issued the following statement:

House of Representatives
Washington, D. C., February 7, 1936
Hon. William P. Connery, Jr.,
Chairman, Committee on Labor,
House of Representatives, Washington, D. C.

My Dear Chairman Connery: In my capacity as a member of the subcommittee which held hearings on the silicosis situation which developed from the construction of the Hawks Nest tunnel at Gauley Bridge, W. Va., I desire to file this individual opinion, to wit:

1. I attended all of the hearings except one, and the testimony tended strongly to impress upon me the belief that the disregard of safety laws and regulations, lack of proper equipment, and unsatisfactory labor conditions contributed in a marked degree to the silicosis problem, which is of national importance.

2. Failure of the officials of the construction firm of Rhinehart [sic] & Dennis to accept the invitation of the committee to appear is regretted, as I am certain the members of the committee desire very strongly to hear the "other side" in this controversy. For this reason I feel that the committee should be give the power of subpoena.

3. Expert medical testimony causes me to believe that a thorough investigation of silicosis throughout the United States would be of much benefit and perhaps bring about a broad understanding of this disease which might well be the basis for constructive State legislation.

*Faithfully yours,
Jennings Randolph* [45]

---

45. "An investigation . . . ." pg. 203.

CHAPTER TWO

INTERVIEWS WITH TUNNEL SURVIVORS

STANLEY CAVENDISH, CHEMIST[46]

Stanley G. Cavendish *was a young chemist employed to work on the Hawks Nest project in 1930. His job entailed monitoring the output from tunnel excavation, taking rock samples as necessary. Cavendish visited underground from time to time, putting him in a good position to observe the work in progress. He recalls construction details clearly today, his view generally reflecting the perspective of company management. This article is edited from a larger 1989 interview by Dennis Deitz, supplemented by a 1990 interview by Ken Sullivan.*

[intro omitted]

Stanley Cavendish: The day they started breaking the face of the mountain up there at Gauley Junction, the original plan was for the tunnel to be 33 feet in diameter. I was there when they first started digging the rock, or getting ready to drill the rock, right at the mouth of where the tunnel was.

They core-drilled a lot of that area [earlier] down 500 feet. Samples were taken, but they were never analyzed. I don't know why. That's a mystery, because they should have known what was underneath the ground. I expect they must have drilled maybe 400 or 500, well, make it 200 different holes all over the area to try to determine the quality of the rock. They had samples that they brought up, but nobody was ever interested in that, except at the time that the tunnel was started.

I took a sample and took it down to the laboratory and found that the rock was almost pure silica—99.9%, along in there somewhere. Very, very, very valuable sandstone. I reported that and the local authorities called New York and told them the situation, and said "I think we've struck a gold mine, because this rock is of extraordinary quality."

As I said, the tunnel was originally planned for 33 feet, 32 or 33 feet in diameter, and in order to get the extrta amount of silica rock out of there, they increased it to 46 feet. That was only at the entrance of the tunnel. Up at the end of the tunnel at the dam site, that was nothing but pure old shale, it was just absolutely worthless and they had to throw it away. The only [valuable] rock coming out of the tunnel was at the face of

---

46. Dennis Deitz. "I Think We've Struck a Gold Mine: A Chemist's View of Hawks Nest," in *Goldenseal*, vol. 16, no. 3. Fall 1990.

Experiments to excavate the tunnel with this machine were unsuccessful. *Courtesy West Virginia State Archives.*

the thing,[47] and that's why it was very important that I monitor that stuff. And after it had run out, the rest of the rock was no good whatsoever.

But they had core-drilled from there clear on up to the dam. They drilled that stuff and withdrew the cores, but I don't think they ever had it analyzed. I don't know why they never did. I've always wondered why that when they got ready to build this thing and they cut the first cuttings where the face of the tunnel was, that we found out that it was such good quality of rock. I know up there at that laboratory they had a lot of core drillings about an inch in diameter. But the quality of the rock, they just never did examine it. I never figured out why it was.

Dennis Dietz: Were you in the tunnel when it was actually being built, when the workers were in there?

SC: Yes, I was in the tunnel, but not too many times. Ocasionally, I would have to go in to check out a sample or something like that. Other than that, I didn't normally go in the tunnel. But I was in it several times and I saw the operation that was going on to build the tunnel. It was a lot of workers they had at the face of the tunnel. They had steps down from the top of the tunnel, the face of the tunnel, down to the bottom

---

47. Cavendish is referring to the tunnel opening at the power house, Gauley Junction.

where their drillers were busy with air drills, drilling holes so that they could put the dynamite in there and shoot.

[omission]

I never wore a mask at all. I wasn't in the tunnel long enough. I didn't realize they were in such a bad situation. Matter of fact, I guess I didn't even know anything about silicosis at all. But I don't think I was in the tunnel long enough to do any detrimental effect to my lungs.

DD: How long was it before they began to suspect that it was affecting the workers?

SC: Well, that's hard to say. I really don't know. I began to hear that the people were affected by the dust. I believe they realized to a certain extent that the dust was detrimental, but the contractors, Rinehart and Dennis out of Charlottesville, Virginia, I'm not sure that they had a whole lot of concern about safety. That might be an opinion of mine, but I don't think they did. They did try to put water on these drills, to keep down the dust. But the drills would jam and they couldn't operate them, so they just abandoned that idea after trying it. It didn't work.

[omission]

DD: The silica rock, sand or whatever, you said it was valuable. For what?

SC: To making ferrosilicon. What they did after the plant was finished and the whole project was completed, when they started operation, they made the ferrosilicon by mixing in steel shavings. They brought steel shavings in cars, coal cars, and dumped them there on the site where the plant was, and in the meantime they had dumped this tremendous, tremendous mountain of silica rock there, and they mixed these two together in an electric furnace. And they made the ferrosilicon which they shipped to steel companies to go into various forms of steel, like stainless steel, something like that. In addition to that, they also made ferromanganese there, too.

[remainder of main article omitted]

This following information was a sidebar within the main article.

The decision to increase the diameter of the tunnel in order to mine the silica rock was a critical point in the Hawks Nest episode, expanding the project and extending the workers' exposure to the deadly dust. As chemist Stanley Cavendish recalls it, the fateful decision was based on his last-minute discovery of the purity of the silica, and not premeditated as critics charged.[48] *Goldenseal* editor Ken Sullivan asked him about it in an interview from June, 1990.

---

48. Without challenging the veracity of Mr. Cavendish, it seems unlikely that Union Carbide would have spent time and money procuring hundreds of core samples and then neglect to ascertain the results.

Courtesy West Virginia State Archives.

Ken Sullivan: Once you made the discovery, how was the decision made to expand the tunnel?

Stanley Cavendish: The people in New York made the decision to go ahead and expand the tunnel and to get this extra amount of silica out that they hadn't planned to do before. I reported my findings to my superior. His name was B. G. Doom, and he was assistant plant manager. He passed it on to the plant manager, and then on up the line, to whoever it was necessary to go to.

KS: Was there any excitement as the decision was made?

SC: No, I don't think so. I don't recall any. It was just a matter of what to do with it, to get all this extra amount of rock. You see, that rock was valuable. They took it out to the [Alloy] plant and stored it down there. I don't know how much they had. They carried it off for a long, long time.

KS: And it was the key ingredient in the Alloy manufacturing process?

SC: Sure, absolutely.

KS: Where had they got their silica before?

SC: I can't tell you that. I don't know. They'd never made it there before. They had a plant in Niagara Falls that made these alloys. Whether they made any raw silica up there or not, I'm not sure. I think they did.

KS: Did they made ferrosilicon products locally after the discovery of silica at the tunnel?

SC: Oh, sure. That was their intentions. Now you see, when they planned their whole project, I assume that they realized that they were going to have to have silica. There's a place up at Thayer, West Virginia, an outcropping of this whole big layer of sandstone. [There used to be] a quarry up there that mined the stuff, and I assumed that would be a source of their material, the raw material. But when they struck this thing, well, then that was a much better situation.

KS: Do you recall how long it was from the time you had reported the purity of the silica until the decision was made to expand?

SC: Less than a day. Less than that. Maybe the same day, I'm not sure. I know it wasn't very long.

B.H. METHENEY[49]

We at Goldenseal *were surprised to discover a Fayette County survivor of the tunnel, very much alive and in reasonably good health a half-century later. This article is edited from a longer interview with Mr. Metheney, taped at his home in Deep Water last spring.*

David Orr: How did you come to work on the Hawk's Nest Tunnel project?

B.H. Metheney: I was working on a lumber job up at Swiss, on Gauley River. It worked out so the tunnel job was just starting and I went down and started on it in March, I reckon it was. I worked on it up until late fall and quit, and went up to Beckley and worked on the road a little while, maybe six weeks or so, and come down here to Alloy and went to work. They were starting to build this powerhouse at Alloy at that time, and a big mud hole there. I went up to work there with it and just stayed there.

When I came to Hawks Nest looking for a job, I guess it was 1930-31, as well as I remember, and they were starting to grade the railroad around the portal there at Gauley Junction. I fired steam shovels while we graded around there, and then after they opened the tunnel I went to run the drill inside, run a drill for a while.

That tunnel was a rough job. They run the grade around there—till they could get the railroad around there, till they could get the railroad around, bring their equipment around, then shot out the portal. They shot that out with black powder. The churn drill there used about six colored men. They drilled holes about 40 feet deep, just pick it up and drop it, six of them lift it up and drop it. They started springing them holes, put just a little powder in and shoot that. Then they'd put in a little bigger charge, and

---

49. David Orr and Jon Dragan. "A Dirty, Messy Place to Work: B.H. Matheney Remembers Hawks Nest Tunnel," in *Goldenseal*, vol. 7., no. 1, January-March 1981.

shoot it again. They kept on doing that, I don't know, maybe one whole shift. And then they just started carrying them cans of black powder, pouring 'em in, I guess it was 50-pound cans. I don't know how many they poured in, but lots of it. And when they set that off it took that whole block out. Some of it went clear across the river to the C&O track.

[omission]

They started that hole 32 feet in diameter, and then after they got under a little ways, they widened it to 56 feet because that rock was silica. They wanted that rock for the furnaces down here. They crushed it and hauled it down here and stored it at Alloy. When they started operating down there, they used it. But now the upper end was a sort of shale, and they just hauled it out and dumped it. It was no good for anything. They had to crib up there to hold the top. But this lower end, when they got started under there, they'd just put off a shot. A couple guys would go in there with bars. Any rock a-hanging, they'd pry 'em down. That's all they had to do. She was solid as could be.

They had four heading there to work from. They worked the lower end called Number One; up at the surge basin they turned both ways, cause Two and Three was there. Then up at Hawks Nest, at the dam, they were driving back down. And that way they were drilling four ways on it.

They worked day and night. I worked in the lower end, down here, and worked on the night shift. We'd get there about 6:00. The other crew had done come out, had their shot put off. Then we'd go in and start drilling on ours, and we'd drill and load, and shoot, and we'd finish about 6:00 the next morning.

They'd run about eight heading drills, that drilled horizontal. And they'd run eight jackhammers, drilled down. The horizontal shot was carried ahead of the rest of them. They drilled some short holes, and then they'd keep getting longer steel and bring 'em out a little bit, see. And they'd shoot that with delayed explosives. They'd set their holes so far apart, and of course it's just a split second, oh, just a small part of a second between explosions, but it was enough so that it would spring 'em out of the way.

Then after that was shot, they worked on the bench. We had beds, and then the lower part of it we called a toe, and we would just keep it on grade, you know, maybe some places drill holes two foot deep, maybe four foot deep, but never more than six foot. But then when you got up on the next level, towards where the heading was, you drilled about 14-foot holes. And you used about 90 percent dynamite to load with. You'd have to load the hole with 90% and then instead of using sand or something to tamp with, they'd take 40 percent dynamite and tamp it. The whole thing was just full of explosives. Getting ready to put it off, you shot it was 110 volt current—they didn't use a battery you know. Just connect it to the light wire and shoot it.

DO: Did you start as a foreman?

BHM: No, I didn't work as a foreman. Oh, I did maybe for a couple of weeks; I was muck foreman. I ran a drill for the most part.

Tunnel portal at power house. Heading is approximately 20 feet. *Courtesy West Virginia State Archives.*

DO: What kind of drill was that?

BHM: Jackhammers like they use outside except they was big—they weighed 120 pounds. They had connections on 'em to put water on 'em so that you could keep the dust down. But they would cut a lot faster running dry, so they run 'em dry except when they knew the inspector was coming. Then you'd put water on them until he looked 'em over and had gone back outside and then you threw the water off again. You'd come out of there of a morning and catch your nose and squeeze it and it's just like squeezing toothpaste. Clear white stuff.

Jon Dragan: Who told you to cut the water off on the drills? Was that just so you could make time, or once the inspecstor [sic] left did they tell you to cut it off?

BHM: The foreman would tell you, "Get the water off of 'em. Get to drilling." But now, the heading drills, they run them wet all the time. The ones that drilled horizontal, because they'd get that dust out of the holes better that way. But the others—when they'd shoot, you know, it would make some cracks in that rock, and if you got water in there, in that dust, you'd hang a steel, have to maybe discard that hole, when you were already down six or eight feet, move over and start another hole. They didn't like it.

Had a big air compressor that set outside that supplied the air. I think it was five-inch line coming in there, and they had a section down the tunnel that they kept a fire built on, to heat that air. And then sometimes, I've had ice to come out of them exhausts from them drills that'd cut the blood out of you. Used 120 pounds of air pressure.

[omission]

JD: Did you drill so many hours a day and then shoot once a day?

BHM: Yeah. Each crew made a cut. You would take the tunnel ahead about six foot. Every time you shot, you'd move ahead about six foot. That meant 12 feet every 24 hours with the two crews.

The heading drills, the ones that drilled horizontal, they used water on them all the time, but the bench drills, they run dry. It was 32 foot wide—I'm talking about Number One heading down here, now. They used 100-watt bulbs, and they was about ten foot apart, strung on a wire. At midnight or two in the morning you couldn't tell whether a man was a white man or a colored man to look at him across that 32 feet. You'd probably be looking 20 feet or something because he wouldn't be against the rib on that side and you wouldn't either, on the other. The dust couldn't have been no worse, or you couldn't have breathed it. You would have just had to chew it, ate it.

For a good while they used little gasoline motors to haul muck cars out with. They would get carbon monoxide fumes off of that, and I've seen men fall out there. I've helped load as many as 15 on them muck cars at a time, and haul 'em out in the fresh air. Pour a little ammonia into them and they come up out of there vomiting—go back and go to work again.

DO: Did you take a break during the shift?

BHM: No, wasn't no breaks. Thirty-minute lunch.

[omission]

DO: Where did they [laborers] live?

BHM: They had a bunch of shacks up where the golf course is now.[50] Rinehart and Dennis set up the shacks—they charged them rent, you know. They lived up there. Some of them had women with them; some was just five or six men in one shack. Some of them in summertime—and of course it was summer about all the time I worked there—would just lay up there in the brush. They'd do their night shift and go up to the commissary and get 'em a bottle of pop and a five-cent cake or something, and eat that and then crawl back in the brush and sleep until time to go back on shift again.

Up there at the shacks they had a shack rouster they called McCloud—was a white man—who came in there with Rinehart and Dennis. He had one shack he let them

---

50. Hawks Nest Country Club, U.S. Rt. 60, Gauley Mountain.

play poker in, and he'd cut the pot on them, you know, get a rake-off there. Well, if they done any drinking there, as long as they was drinking out of a Coca-Cola bottle, it was all right. It was $2.00 for a Coca-Cola bottle of moonshine. But if they caught them with anything else, why, he'd take 'em and turn 'em over [to] the deputy sheriff. They'd take 'em up to Fayetteville. So that was his rake-off. He was bootlegging in Coca-Cola bottles and if they was drinking his liquor, why it was all right to drink.

[omission]

There was a concrete inspector, tester. He tested to see if it was up to strength. Name was Patterson, Andy Patterson. And he got all this moonshine. It must have been pretty rotten, because he couldn't drink it. He gave it to this old nipper, and I guess he got pretty high on it 'cause he missed a shift of work. Patterson asked him, "How was that liquor I give you?" He said, "Just right." "What do you mean, 'just right'?" He said, "If it had been any worse, I couldn't have drunk it. If it had been any better, you wouldn't have give it to me!"

DO: How many workers were there, do you know?

BHM: I wouldn't have the least idea how many there were. More than a hundred, but I don't know how many. They came and went all the time. You couldn't tell who was going where, you know. On the crew I worked on, down at the lower end there, I'd say there was the drillers, the muckers—all told, maybe 60 men to the shift. That's just a guess. I never did count 'em.

[omission]

DO: Did they work two shifts?

BHM: Oh, yeah, they worked two shifts. Well, the way it was, each crew was supposed to take out a cut. If you got done in nine or ten hours you still got that shift for it, 12 hours. But if it took you 13 hours you still done it. And that held the other crew up and they had to work like the devil to get back on schedule if you were behind. It took about all of your 12 hours, usually, to get it done.

[omission]

DO: Weren't any people coming in distributing stuff, or asking about things? How working conditions were?

BHM: No. I never saw any of them around. One guy came in there with respirators one time and tried to sell them and the company wouldn't buy them. If we had had them they would have helped, but we never got to try them. Anything thick enough, even a handkerchief over your face, would have helped.

DO: Did the workers try that?

BHM: No. No, I don't know. I guess all the rest were like me, it was just a dirty messy place to work. I never thought about the health angle of it.

Power House Camp Site No. 2. Contractor's main office in center. *Courtesy West Virginia State Archives.*

[omission]

DO: Did you consider it a good job or bad? How do you remember it?

BHM: I remember it as a bad job. But at the time, a job was a job. That was the only thing there was in this whole country. When you got some babies looking at you for something to eat you're going to work.

C. LOUISE HARLESS[51]

Back then, back in the 30s, Gauley Bridge was real different than now. I'm guessin' the population was maybe four or five hundred people. But it was a busy little place. Specially on a Saturday. Old Harvey Layne had a garage and there was Bracken's bus station. Greyhound station is what it was. I worked there one time. Up the street was Wiseman's piece goods store. Had our own 10-cent store and a theatre. Sheriff Conley had the theatre.

---

51. From author's interview with Louise Harless, November, 2003. Louise Harless was married to Dr. Harless' youngest son, Walter.

We didn't have a mayor 'til much later, so I guess you could say the local politicians ran the town. Make that singular. Sheriff Conley. (laughter) He was fine if he liked you but if he didn't, oooh boy, you'd better watch out. Walter and I rented from him and Ms. Conley when we first got married and I got along with him just fine.

Rentin's one way Sheriff made his money. Made the rest of it like most of 'em then—bootlegging. I drank it once or twice and that was the end of it cause it'd kill ya if you drank too much of it. Oh, lord. Well, all of us did it to tell you the truth.

I was only nineteen when I got married. Too young to vote. But I saw Herbert Hoover. He came through Gauley Bridge on the back of a train.

People these days can't visualize what that those times were like—the depression and all. Folks were *hungry*. Why I can remember lots of times when Walter and I would pick blackberries because we didn't have anything else to eat. Just go out and pick 'em and eat 'em. Uncle Bob used to call me "bird legs."

It's hard to explain those times. The theatre showed pictures and things about it before the main feature. Besides listening to the radio, that's how we got the news. Pathé News they called it. Showed all the suffering and hungry folk. I guess they wanted the rest of the world to know how bad it got. But I'll tell you this—there was a lot of money made too. Seems like that's always been the thing.

That's just about the time they started digging the tunnel. Walter worked at times on the river doing "soundings." That's before the tunnel work got started. About the time those men drowned up there. A sad, sad day that was.

Doc Harless wouldn't let Walter go to work in the tunnel, that's for sure. He knew from the start something was wrong. The merchants and townspeople didn't want to hear anything about the dying. They didn't think it was true, I guess. And I'll tell you something else. Union Carbide and that construction company knew something was wrong. They knew what was happening with those men. Doc Harless tried to tell them but they were making too much money. That was the bottom line

Doc Harless had already been fired as company doctor when this all started. Witt Jennings was head of the "M" company[52] then. One day Doc Harless had a colored lady in labor the same time another lady, a white lady in Glen Ferris, went into labor. The "M" people told him to leave the colored lady and go to the one in Glen Ferris. He wouldn't do it. They said he had to. Doc Harless went as soon as he could but he would not leave that other woman. That's the reason he was fired. After that he set up his own practice in Gauley Bridge, a private practice. Those men from the tunnel would come to see him, but he wasn't the company doctor after that.

Doc Harless had a mind of his own, you'd sure have to say that, and he let you know what he was thinking. When Mom Harless got sick, we went to live with them. At first I cried all the time cause he'd say something hateful to me. But he found me one

---

52. Local vernacular for the Electrometallurgical Company.

day and I wasn't feeling like hearing what he had to say, so I dished it right back out to him. He laughed and laughed at what I said and from then on I got along just fine with him.

He had a hard time getting along with people—Doc Harless did. He was abrupt. Very abrupt and if you didn't know him you just wouldn't like him, now that's the truth. But for his time and medicine, he was a good doctor, and really he was a good man. Had a good heart. But as for being a man that was sympathetic and all, he wasn't. He just was his own self and didn't anybody tell him what to do. That's the reason he walked out of that trial in Fayetteville. He told 'em to go to hell and got up and walked out. Said it just wasn't right—the way those men were treated.

Those people came out of the deep, deep south. The blacks, that is. There were so many of them and I don't think they could even read or write. When they got here they'd stay any place they could find, trying to make some money. Weren't paid anything to amount to at the tunnel, I know that.

Always has been discrimination, I'd say, but they sure were treated bad.

If they wanted something from a diner, they ordered 'round at the back and then the food was handed out to them like they were a dog on the street. Couldn't go inside. Had separate water fountains, too. Just simply weren't treated right.

There's another thing. When those men got so sick they couldn't work and just wanted to go back to where they came from—this was after they got that little bit of nothing from the trial and paid off the lawyers—what little bit they had left they used to buy old cars. I doubt if they ever got out of Gauley Bridge. He sold them old cars that wouldn't go a lick—Layne did. It goes right back to money again.

I can remember seeing one or two of those men. You couldn't tell by looking except they looked like someone who was sick. Vic[53] and Walter both used to haul them to Charleston for x-rays. Doc Harless didn't have an x-ray machine in his office, but he kept those x-rays.

And he kept the lungs too. The ones from the autopsies. They were in big jars. Weighed about eight pounds, I'd say. That's when we lived with him up on the hill in Gauley Bridge. He brought them home and put them down in the basement. There was a little place down there where we stored things. Canned things and all.

I had this colored woman come in one day to help me. She came up stairs where I was and said, "Ms. Harless, what's that in those jars?" Like a fool I told her what they were and she never came back again, so help me.

Years later when we moved, I told Walter, "I'm not carrying those things with me." He said we'd put them in the back of the truck and take them up on Cotton Hill Mountain—on the old road to Beckwith. They sure made a racket when they went down over the mountain but it sounded like just one jar broke. I didn't know what

---

53. Vic Harless, Doc Harless' eldest son.

else to do with the damn things. Walter and I called them the "Jones Boys." Just made it seem less awful that way.

CHAPTER THREE

LITERATURE SPAWNED BY THE INCIDENT AT HAWKS NEST

U. S. 1, BOOK OF THE DEAD
Muriel Rukeyser

STATEMENT: PHILIPPA ALLEN

You like the State of West Virginia very much do you not?
I do very much, in the summertime.
—How much time have you spent in West Virginia?
—During the summer of 1934, when I was doing social work
down there, I first heard of what we were pleased to call
the Gauley tunnel tragedy, which involves about 2,000
men.
—What was their salary?
—It started at 40¢ and dropped to 25¢ an hour.
—You have met these people personally?
—I have talked to people: yes.
According to estimates of contractors
2,000 men were
      employed there
      period, about 2 years
      drilling, 3.75 miles of tunnel.
      To divert water (from New River)
      To a hydroelectric plant (at Gauley Junction).
The rock through which they were boring was of a high
      silica content.
In tunnel No. 1 it ran 97-99% pure silica.
The contractors
      Knowing pure silica
      30 years' experience
      must have known danger for every man
neglected to provide the workmen with any safety device....

—As a matter of fact, they originally intended to dig that
    tunnel a certain size?
—Yes.
—And then enlarged the size of the tunnel, due to the fact
    that they discovered silica and wanted to get it out?
—That is true for tunnel No. 1.
The tunnel is part of a huge water-power project
    begun, latter part of 1929
        direction: New Kanawha Power Co.
         subsidiary of Union Carbide & Carbon Co.
        That company—licensed:
        To develop power for public sale.
        Ostensibly it was to do that; but
        (in reality) it was formed to sell all the power to
            the Electro-Metallurgical Co.
        subsidiary of Union Carbide & Carbon Co.
                which by an act of the State legislature
                was allowed to buy up
        New Kanawha Power Co. in 1933.
—They were developing the power. What I am trying to
get at, Miss Allen, is, did they use this silica from the
tunnel; did they afterward sell it and use it in commerce?
—They used it in the electro-processing of steel.
SiO_2 SiO_2
The richest deposit.
Shipped on the C&O down to Alloy.
It was so pure that
                SiO_2
They used it without refining.
—Where did you stay?
—I stayed at Cedar Grove. Some days I would have to hitch
into Charleston, other days to Gauley Bridge.
—You found the people of West Virginia very happy to pick
you up on the highway, did you not?
—Yes: they are delightfully obliging.
(All were bewildered. Again at Vanetta they are asking,
"What can be done about this?")
I feel that this investigation may help in some matter.
I do hope it may.
I am now making a very general statement as a beginning.
There are many points that I should like to develop

later, but I shall try to give you a general history of
this condition first . . . .

PRAISE FOR THE COMMITTEE

These are the lines on which a committee is formed.
>Almost as soon as work was begun in the tunnel
>men began to die among dry drills. No masks.
>Most of them were not from this valley.
>The freights brought many every day from States
>all up and down the Atlantic seaboard
>and as far inland as Kentucky, Ohio.
>After the work the camps were closed or burned.
>The ambulance was going day and night,
>White's undertaking business thriving and
>his mother's cornfield put to a new use.
>"Many of the shareholders at this meeting

"were nervous about the division of the profits;
"How much has the Company spent on lawsuits?"
"The man said $150,000. Special counsel:
"I am familiar with the case. Not: one: cent.
" 'Terms of the contract. Master liable.'
"No reply. Great corporation disowning men who
>made . . . ."

After the lawsuits had been instituted. . . .

The Committee is a true reflection of the will of the people.
>Every man is ill. The women are not affected,
>This is not a contagious disease. A medical commission,
>Dr. Hughes, Dr. Hayhurst examined the chest
>of Raymond Johnson, and Dr. Harless, a former
>company doctor. But he saw too many die,
>he has written his letter to Washington.

The Committee meets regularly, wherever it can.
>Here are Mrs. Jones, three lost sons, husband sick,
>Mrs. Leek, cook for the bus cafeteria,
>the men : George Robinson, leader and voice,
>four other Negroes (three drills, one camp-boy)
>Blankenship, the thin friendly man, Peyton the engineer,
>Juanita absent, the one outsider member.
>Here is the noise, loud belts of the shoe-repair shop,

        meeting around the stove beneath the one bulb hanging.
        They come late in the day. Many come with them
        who pack the hall, wait in the thorough dark.
This is a defense committee. Unfinished business:
        Two rounds of lawsuits, 200 cases
        Now as to the crooked lawyers
        If the men had worn masks, their use would have involved
        time every hour to wash the sponge at mouth.
        Tunnel, 3 ⅛ miles long. Much larger than
        the Holland Tunnel or Pittsburgh's Liberty Tubes.
        Total cost, say, $16,000,000.
This is the procedure of such a committee:
        To consider the bill before the Senate.
        To discuss relief.
                Active members may be cut off relief,
                    16-mile walk to Fayetteville for cheque—
                    WEST VIRGINIA RELIEF ADMINISTRATION, #22991

TO JOE HENIGAN, GAULEY BRIDGE, ONE AND 50/100 WINONA NATIONAL BANK. PAID FROM STATE FUNDS.

Unless the Defense Committee acts;
the **People's Press**, supporting this fight,
signed editorials, sent in funds.
Clothing for tunnel-workers.
        Rumored, that in the post-office
        parcels are intercepted.
        Suspected : Conley. Sheriff, hotelman
        head of the town ring—
        Company whispers. Spies,
        The Racket.
Resolved, resolved.
George Robinson holds all their strength together:
To fight the companies to make somehow a future.

"At any rate, it is inadvisable to keep a community of dying
persons intact."
"Senator Holt. Yes. This is the most barbarous example of
industrial construction that ever happened in the world."
Please proceed.

"In a very general way Hippocrates' Epidemics speaks
    of the metal digger who breathes with difficulty,
    having a pain and wan complexion.
    Pliny, the elder. . . ."
"Present work of the Bureau of Mines. . . ."
The dam's pure crystal slants upon the river.
    A dark and noisy room, frozen two feet from stove.
    The cough of habit. The sound of men in the hall
    waiting for word.
        These men breathe hard
        but the committee has a voice of steel.
        One climbs the hill on canes.
        They have broken the hills and cracked the riches wide.

        In this man's face
        family leans out from two worlds of graves---
        here is a room of eyes,
        a single force looks out, reading our life.

        Who stands over the river?
        Whose feet go running in these rigid hills?
        Who comes, warning the night,
        shouting and young to waken our eyes?

        Who runs through electric wires?
        Who speaks down every road?
        Their hands touched mastery; now they
        demand an answer.

TUNNELITIS

hold themselves up
at the side of a tree.
I can go right now
to that cemetery.

When the blast went off the boss would call out, Come, let's
    go back.
when that heavy loaded blast went white, Come, let's go back,
telling us hurry, hurry, into the falling rocks and muck.

The water they would bring had dust in it, our drinking
        water,
the camps and their groves were colored with the dust,
we cleaned our clothes in the groves, but we always had
        the dust.

Looked like somebody sprinkled flour all over the parks
        and groves,
it stayed and the rain couldn't wash it away and it twinkled
that white dust really looked pretty down around our ankles.

As dark as I am, when I came out at morning after the
        tunnel at night,
with a white man, nobody could have told which man was
        white.
The dust had covered us both, and the dust was white.

JUANITA TINSLEY

Even after the letters, there is work,
sweaters, the food, the shoes
and afternoon's quick dark

draws on the windowpane
my face, the shadowed hair,
the scattered papers fade.

Slow letters! I shall be
always—the stranger said
"To live stronger and free."

I know in America there are songs,
forgetful ballands to be sung,
but at home I see this wrong.

When I see my family house,
the gay gorge, the picture-books,
they raise the face of General Wise

aged by enemies, like faces

the stranger showed me in the town.
I saw that plain, and saw my place.

The scene of hope's ahead: look, April,
and next month with a softer wind,
maybe they'll rest upon their land,
and then maybe the happy song, and love,
a tall boy who was never in a tunnel.

THE DOCTORS

—tell the jury your name.
—Emory R. Hayhurst.
—State your education, Doctor, if you will
Don't be modest about it; just tell about it.

High school Chicago 1899
Univ. of Illinois 1903
M.A. 1905, thesis on respiration
P&S Chicago 1908
2 years' hospital training;
at Rush on occupational disease
director of clinic 2 ½ years.
Ph.D. Chicago 1916
Ohio Dept. of Health, 20 years as
consultant in occupational diseases,
Hygienist, U.S. Public Health Service
and Bureau of Mines
and Bureau of Standards

Danger begins at 25%
Here was pure danger
Dept. of Mines
Came in, was kept away.
  Miner's phthisis, fibroid phthisis,
  Grinder's rot, potter's rot,
  Whatever it used to be called
  These men did not need to die.
—Is silicosis an occupational disease?
—It is.

—Did anyone show you the lungs of Cecil Jones?
—Yes, sir.
—What was that?
—It was Dr. Harless.

"We talked to Dr. L. R. Harless, who had handled many of the cases, more than any other doctor there. At first Dr. Harless did not like to talk about the matter. He said he had been subjected to so much publicity. It appeared that the doctor thought he had been involved in too many of the court cases; but finally he opened up and told us about the matter."
—Did he impress you as one who thought this was a very serious thing in that section of the country?
"Yes, he did. I would say that Dr. Harless has probably become very self-conscious about this matter. I cannot say that he has retracted what he told me, but possibly he had been thrust into the limelight so much that he is more conservative now than when the matter was simply something of local interest."

Dear Sir: Due to illness of my wife and urgent professional duties, I am unable to appear as per your telegram.
> Situation exaggerated. Here are facts:
> We examined. 13 dead. 139 had some lung damage.
> 2 have died since, making 15 deaths.
> Press says 476 dead, 2,000 affected and doomed.
> I am at a loss to know where those figures were
>   obtained.
> At this time, only a few cases here,
> And these only moderately affected.
> Last death occurred November, 1934.

It has been said that none of the men knew of the hazard connected with the work. This is not correct. Shortly after the work began many of these workers came to me complaining of chest conditions and I warned many of them of the dust hazard and advised them that continued work under these conditions would result in serious lung disease. Disregarding this warning many of the men continued at this work and later brought suit against their employer for damages.

While I am sure that many of these suits were based on meritorious grounds, I am also convinced that many others took advantage of this situation and made out of it nothing less than a racket.
In this letter I have endeavored to give you the facts which came under my observation. . . .
If I can supply further information. . . .

Mr. MARCANTONIO: A man may be examined a year after he has worked in a tunnel and not show a sign of silicosis, and yet the silicosis may develop later; is not that true?
—Yes, it may develop as many as ten years after.
Mr. MARCANTONIO: Even basing the statement on the figures, the doctor's claim that this is a racket is not justified?
—No; it would not seem to be justified.
Mr. MARCANTONIO: I should like to point out that Dr. Harless contradicts his "exaggeration" when he volunteers the following: "I warned many. . . ."
(Mr. PEYTON: I do not know. Nobody knew the danger around there.)

Dr. Goldwater. First are the factors involving the individual.
    Under the heading B, external causes.
    Some of the factors which I have in mind—
    those are the facts upon the blackboard,
    the influencing and controlling factors.
Mr. Mr. MARCANTONIO: Those factors would bring about acute
    silicosis?
Dr. Goldwater. I hope you are not provoked when I say
    "might."
    Medicine has no hundred percent.
    We speak of possibilities, have opinions.
Mr. GRISWOLD: Doctors testify answering "yes" and "no."
    Don't they?
Dr. Goldwater. Not by the choice of the doctor.
Mr. GRISWOLD: But that is usual, isn't it?
Dr. Goldwater. They do not like to do that.
    A man with a scientific point of view—
    unfortunately there are doctors without
    that—
    I do not mean to say all doctors are angels—

> but most doctors avoid dogmatic statements.
> avoid assiduously "always," "never."

Mr. GRISWOLD: Best doctor I ever knew said "no" and "yes."
Dr. Goldwater. There are different opinions on that, too.
> We are talking about acute silicosis.
> The man in the white coat is the man on the hill,
> the man with the clean hands is the man with the drill,
> the man who answers "yes" lies still.

—Did you make an examination of those sets of lungs?
—I did.
—I wish you would tell the jury whether or not those lungs were silicotic.
—We object.
—Objection overruled.
—They were.
Stands bare against a line of farther field,
unmarked except for the wood stakes, charred at tip,
few scratched and named (pencil or nail).
Washed-off. Under the mounts,
all the anonymous.
Abel America, calling from under the corn,
Earth, uncover my blood!
Did the undertaker know the man was married?
Uncover.
Do they seem to fear death?
Contemplate.
Does Mellon's ghost walk, povertied at last,
walking in furrows of corn, still sowing,
do apparitions come?
Voyage.
Think of your gardens. But here is corn to keep.
Marked pointed sticks to name the crop beneath.
Sowing is over, harvest is coming ripe.

—No, sir; they want to go on.
They want to live as long as they can.

ARTHUR PEYTON

Consumed. Eaten away. And love across the street.
I had a letter in the mail this morning

Dear Sir, . . .pleasure . . . enclosing herewith our check . . .
payable to you, for $21.59
>being one-half of the residue which
>we were able to collect in your behalf
>in regard to the above case.
In winding up the various suits,
>after collecting all we could,
>we find this balance due you.
With regards, we are
>Very truly,

After collecting
>>the dust the failure the engineering corps
O love consumed eaten away the foreman laughed
they wet the drills when the inspectors came
the moon blows glassy over our native river.

O love tell the committee that I know:
never repeat you mean to marry me.
In mines, the fans are large (2,000 men unmasked)
before his verdict the doctor asked me How long

THE BILL

The subcommittee submits:
Your committee held hearings, heard many witnesses; finds:

THAT the Hawk's Nest tunnel was constructed
>Dennis and Rinehart, Charlottesville, Va., for
>New Kanawha Power Co., subsidiary of
>Union Carbide & Carbon Co.

THAT a tunnel was drilled
>app. dist. 3.75 mis.
>to divert water (from New River)
>to hydroelectric plant (Gauley Junction).

THAT in most of the tunnel, drilled rock contained
>90 –even 99 percent pure silica.

This is a fact that was known.

THAT silica is dangerous to lungs of human beings.
   When submitted to contact. Silicosis

THAT the effects are well known.
   Disease incurable.
   Physical incapacity, cases fatal.
THAT the Bureau of Mines has warned for twenty years.

THAT prevention is: wet drilling, ventilation,
     Respirators, vacuum drills.
  Disregard: utter. Dust : collected. Visibility : low.
  Workmen left work, white with dust.
  Air system : inadequate.
  It was quite cloudy in there.
  When the drills were going, in all the smoke and
    dust,
  It seemed like a gang of airplanes going through
    that tunnel.
  Respirators, not furnished.
  I have seen men with masks, but simply on their
    breasts.
  I have seen two wear them
  Drills: dry drilling, for speed, for saving.
  A fellow could drill three holes dry for one hole
    wet.
  They were so fast they didn't square at the top.
  Locomotives: gasoline. Suffering from monoxide
    gas.
  There have been men that fell in the tunnel. They
    had to be carried out.
The driving of the tunnel.
  It was begun, continued, completed, with gravest
    disregard.
  And the employees? Their health, lives, future?
Results and infection.
  Many died. Many are not yet dead.
  Of negligence. Willful or inexcusable.
Further findings:
  Prevalence: many States, mine, tunnel operations.
  A great menace.

We suggest hearings be read.
> This is the dark. Lights strung up all the way.
> Depression; and, driven deeper in,
> by hunger, pistols, and despair,
> they took the tunnel.

Of the contracting firm
> P.H. Faulconer, Pres.
> E.J. Perkins, Vice-Pres.
> have declined to appear.
> They have no knowledge of deaths from silicosis.
> However, their firm paid claims.
> I want to point out that under the statute $500 or
> > $1000, but no more, may be recovered.

We recommend.
> Bring them. Their books and records.
> Investigate. Require

Can do no more.
> These citizens from many States
> paying the price for electric power,
> To Be Vindicated.

"If by their suffering and death they will have made a
future life safer for work beneath the earth, if they will have
been able to establish a new and greater regard for human
life in industry, their suffering may not have been in vain."
> Respectfully,
> Glenn Griswold
> Chairman, Subcommittee
> > Vito Marcantonio
> > W.P. Lambertson
> > Matthew A. Dunn

The subcommittee subcommits.

Words on a monument.
Capitoline thunder. It cannot be enough.
The origin of storms is not in clouds,
our lightning strikes when the earth rises,
spillways free authentic power:
dead John Brown's body walking from a tunnel
to break the armored and concluded mind.

MAN ON A ROAD
Albert Maltz

At about four in the afternoon I crossed the bridge at Gauley, West Virginia, and turned the sharp curve into the tunnel under the railroad bridge. I have been over this road once before and knew what to expect—by the time I entered the tunnel I had my car down to about ten miles an hour. But even at that speed I came closer to running a man down than I ever have before. This is how it happened.

The patched, macadam road had been soaked through by an all-day rain and now was as slick as ice. In addition, it was pitch dark—a black sky and a steady, swishing rain made driving impossible without headlights. As I entered the tunnel a big truck swung fast around the curve on the other side. The curve was so sharp that his headlights had given me no warning. The tunnel was short and narrow, with about passing space for two cars, and before I knew it he was in front of me with his big, front wheels over on my side of the road.

I jammed on my brakes. Even at ten miles an hour my car skidded, first toward the truck and then, as I wrenched on the wheel, in toward the wall. There it stalled. The truck swung around hard, scraped my fender and passed through the tunnel about an inch away from me. I could see the tense face of the young driver with the tight bulge of tobacco in his cheek and his eyes glued on the road. I remember saying to myself I hoped he'd swallow that tobacco and go choke himself.

I started my car and shifted into first. It was then I saw for the first time that a man was standing in front of my car about a foot away from the inside wheel. It was a shock to see him there. "For Chrissakes," I said.

My first thought was that he had walked into the tunnel after my car had stalled. I was certain he hadn't been in there before. Then I noticed that he was standing profile to me with his hand held up in the hitch-hiker's gesture. If he had walked into that tunnel, he'd be facing me—he wouldn't be standing sideways looking at the opposite wall. Obviously I had just missed knocking him down and obviously he didn't know it. He didn't even know I was there.

It made me run weak inside. I had a picture of a man lying crushed under a wheel with me standing over him knowing it was my car.

I called out to him "Hey!" He didn't answer me. I called louder. He didn't even turn his head. He stood there, fixed, his hand up in the air, his thumb jutting out. It scared me. It was like a story by Bierce where the ghost of a man pops out of the air to take up his lonely post on a dark country road.

My horn is a good, loud, raucous one and I knew that the tunnel would re-double the sound. I slapped my hand down on that little black button and pressed as hard as I could. That man was either going to jump or else prove that he was a ghost.

Well, he wasn't a ghost—but he didn't jump, either. And it wasn't because he was deaf. He heard that horn all right.

He was like a man in a deep sleep. The horn seemed to awaken him only by degrees, as though his whole consciousness had been sunk in some deep recess within himself. He turned his head slowly and looked at me. He was a big man, about thirty-five with a heavy-featured face—an ordinary face with a big, fleshy nose and a large mouth. The face didn't say much. I wouldn't have called it kind or brutal or intelligent or stupid. It was just the face of a big man, wet with rain, looking at me with eyes that seemed to have a glaze over them. Except for the eyes you see faces like that going into the pit at six in the morning or coming out of a steel mill or foundry where heavy work is done. I couldn't understand that glazed quality in his eyes. It wasn't the glass stare of a drunken man or the wild, mad glare I saw once in the eyes of a woman in a fit of violence. I could only think of a man I once knew who had died of cancer. Over his eyes in the last days there was the same dull glaze, a far away, absent look as though behind the blank, outward film there was a secret flow of past events on which his mind was focused. It was this same look that I saw in the man on the road.

When at last he heard my horn, the man stepped very deliberately around the front of my car and came toward the inside door. The least I expected was that he would show surprise at an auto so dangerously close to him. But there was no emotion to him whatsoever. He walked slowly, deliberately, as though he had been expecting me and then bent his head down to see under the top of my car. "Kin yuh give me a lift, friend?" he asked me.

I saw his big horse teeth chipped at the ends and stained brown by tobacco. His voice was high-pitched and nasal with the slowed, lilting drawl of the deep South. In West Virginia few of the town folk seem to speak that way. I judged he had been raised in the mountains.

I looked at his clothes—an old cap, a new blue work shirt and dark trousers, all soaked through with rain. They didn't tell me much.

I must have been occupied with my thoughts about him for some time, because he asked me again. "Ahm goin' to Weston," he said. "Are you a-goin' thataway?"

As he said this, I looked into his eyes. The glaze had disappeared and now they were just ordinary eyes, brown and moist.

I didn't know what to reply. I didn't really want to take him in—the episode had unnerved me and I wanted to get away from the tunnel and from him too. But I saw him looking at me with a patient, almost humble glance. The rain was streaked on his face and he stood there asking for a ride and waiting in simple concentration for my answer. I was ashamed to tell him "no." Besides, I was curious. "Climb in," I said.

He sat down beside me, placing a brown paper package on his lap. We started out of the tunnel.

From Gauley to Weston is about a hundred miles of as difficult mountain driving as I know—a five mile climb to the top of a hill, then five miles down and then up another. The road twists like a snake on the run and for a good deal of it there is a jagged cliff on one side and a drop of a thousand feet or more on the other. The rain and the small rocks crumbling from the mountain sides and littering up the road made it very slow going. But in the four hours or so that it took for the trip I don't think my companion spoke to me half a dozen times.

I tried often to get him to talk. It was not that he wouldn't talk, it was rather that he didn't seem to hear me—as though as soon as he had spoken, he would slip down into that deep, secret recess within himself. He sat like a man dulled by morphine. My conversation, the rattle of the old car, the steady pour of rain were all a distant buzz—the meaningless, outside world that could not quite pierce the shell in which he seemed to be living.

As soon as we had started, I asked him how long he had been in the tunnel.

"Ah don' know," he replied. "A good tahm, ah reckon."

"What were you standing there for—to keep out of the rain?"

He didn't answer. I asked him again, speaking very loudly. He turned his head to me. "Excuse me, friend," he said, "did you say somethin'?"

"Yes," I answered. "Do you know I almost ran you over back in the tunnel?"

"No-o," he said. He spoke the work in that breathy way that is typical of mountain speech.

"Didn't you hear me yell to you?"

"No-o." He paused. "Ah reckon ah was thinkin'."

"Ah reckon you were," I thought to myself. "What's the matter, are you hard of hearing?" I asked him.

"No-o," he said, and turned his head away looking out front at the road.

I kept right after him. I didn't want him to go off again. I wanted somehow to get him to talk.

"Looking for work?"

"Yessuh."

He seemed to speak with an effort. It was not a difficulty of speech, it was something behind, in his mind, in his will to speak. It was as though he couldn't keep touch between his world and mine. Yet when he did answer me, he spoke directly and coherently. I didn't know what to make of it. When he first came into the car I had been a little frightened. Now I only felt terribly curious and a little sorry.

"Do you have a trade?" I was glad to come to that question. You know a good deal about a man when you know what kind of work he follows and it always leads to further conversation.

"Ah ginerally follows the mines," he said.

"Now," I thought, "we're getting somewhere,"

But just then we hit a stretch of unpaved road where the mud was thick and the ruts were hard to follow. I had to stop talking and watch what I was doing. And when we came to paved road again, I had lost him.

I tried again to make him talk. It was no use. He didn't even hear me. Then finally, his silence shamed me. He was a man lost somewhere within his own soul only asking to be left alone. I felt wrong to keep thrusting at his privacy.

So for about four hours we drove in silence. For me those hours were almost unendurable. I have never seen such rigidity in a human being. He sat straight up in the car, his outward eye fixed on the road in front, his inward eye seeing nothing. He didn't know he was in the car, he didn't know he was in the——.

He didn't feel the rain that kept sloshing in on him through the rent in the side curtains. He sat like a slab of moulded brick and only from his breathing could I be sure that he was alive. His breathing was heavy.

Only once in that long trip did he change a posture. That was when he was seized with a fit of coughing. It was a fierce, hacking cough that shook his big body from side to side and doubled him over like a child with the whooping cough. He was trying to cough something up—I could hear the phlegm in his chest—but he couldn't succeed. Inside him there was an ugly, scraping sound as though cold metal were being rubbed on the bone of his ribs, and he kept spitting and shaking his head.

It took almost three minutes for the fit to subside. The he turned around to me and said, "Excuse me, friend." That was all. He was quiet again.

I felt awful. There were times when I wanted to stop the car and tell him to get out. I made a dozen good excuses for cutting the trip short. But I couldn't do it. I was consumed by a curiosity to know what was wrong with the man. I hoped that before we parted, perhaps even as he got out of the car, he would tell me what it was or say something that would give me a clue.

I thought of the cough and wondered if it were T.B. I thought of cases of sleeping sickness I had seen and of a boxer who was punch drunk. But none of these things seemed to fit. Nothing physical seemed to explain this dark, terrible silence, this intense, all exclusive absorption within himself.

Hour after hour of rain and darkness!

Once we passed the slate dump of a mine. The rain had made the surface burst into flame and the blue and red patches flickering in a kind of witch glow on a hill of black seemed to attract my companion. He turned his head to look at it, but he didn't speak, and I said nothing.

And again the silence and rain! Occasionally a mine tipple with the cold, dreary, smoke smell of the dump and the oil lamps in the broken down shacks where the miners live. Then the black road again and the shapeless bulk of the mountains.

We reached Weston at about eight o'clock. I was tired and chilled and hungry. I stopped in front of a café and turned to the man.

"Ah reckon this is hit," he said.

"Yes," I answered. I was surprised. I had not expected him to know that we had arrived. Then I tried a final plunge. "Will you have a cup of coffee with me?"

"Yes," he replied, "thank you, friend."

The "thank you" told me a lot. I knew from the way he said it that he wanted the coffee but couldn't pay for it; that he had taken my offer to be one of hospitality and was grateful. I was happy I had asked him.

We went inside. For the first time since I had come upon him in the tunnel he seemed human. He didn't talk, but he didn't slip inside himself either. He just sat down at the counter and waited for his coffee. When it came, he drank it slowly, holding the cup in both hands as though to warm them.

When he had finished, I asked him if he wouldn't like a sandwich. He turned around to me and smiled. It was a very gentle, a very patient smile. His big, lumpy face seemed to light up with it and become understanding and sweet and gentle.

The smile shook me all through. It didn't warm me—it made me feel sick inside. It was like watching a corpse begin to stir. I wanted to cry out "My God, you poor man!"

Then he spoke to me. His face retained that smile and I could see the big, horse teeth stained by tobacco.

"You've bin right nice to me, friend, an' ah do appreciate it."

"That's all right," I mumbled.

He kept looking at me. I knew he was going to say something else and I was afraid of it.

"Would yuh do me a faveh?"

"Yes," I said.

He spoke softly. "Ah've got a letter here that ah done writ to mah woman, but ah can't write very good. Would you all be kind enough to write it ovah for me so it'd be proper like?"

"Yes," I said, "I'd be glad to."

"Ah can tell you all know how to write real well," he said, and smiled.

"Yes."

He opened his blue shirt. Under his thick woolen underwear there was a sheet of paper fastened by a safety pin. He handed it to me. It was moist and warm and the damp odor of wet cloth and the slightly sour odor of his flesh clung to it.

I asked the counterman for a sheet of paper. He brought me one. This is the letter I copied. I put it down here in his own script.

> *My dere wife—*
> *i am awritin this yere leta to tell you somethin i did not tell you afore i lef frum home. There is a cause to wy i am not able to get me any job at the mines. i told you hit was frum work abein slack. But this haint so.*
>
> *Hit comes frum the time the mine was shut down an i worked in the tunnel nere Gauley Bridge where the govinment is turning the river inside the mounten. The mine supers say they wont hire any men war worked in thet tunnel.*
>
> *Hit all comes frum thet rock thet we all had to drill. Thet rock was silica and hit was most all of hit glass. The powder frum this glass has got into the lungs of all the men war worked in thet tunnel thru their breathin. And this has given to all of us a sickness. The doctors writ it down for me. Hit is silicosis. Hit makes the lungs*

to git all scab like and then it stops the breathin.

Bein as our hom is a good peece frum town you aint heerd about Tom Prescott and Hansy McCulloh having died two days back. But wen i heerd this i went to see the doctor.

The doctor says i hev got me thet sickness like Tom Prescott and thet is the reason wy i am coughin sometime. My lungs is agittin scab like. There is in all ova a hondred men war have this death sickness frum the tunnel. It is a turible plague becus the doctor says this wud not be so if the company had gave us masks to ware and put a right fan sistem in the tunnel.

So i am agoin away becus the doctor says i will be dead in about fore months.

i figger on getting some work maybe in other parts. i will send you all my money till i caint work no mohr.

i did not want i should be a burdin upon you all at hum. So thet is wy i hev gone away.

i think wen you daon here frum me no mohr you orter go to your grandmaws up in the mountens at Kilney Run. You kin live there and she will take keer of you an the young one.

i hope you will be well and keep the young one out of the mines. Doan let him work there.

Doan think hard on me for agoin away and doan feel bad. But wen the young one is agrowed up you tell him wat the company has done to me.

i reckon after a bit you shud try to git you anotha man. You are a young woman yit.

Your loving husband,
Jack Pitckett.

When I handed him the copy of his letter, he read it over. It took him a long time. Finally he folded it up and pinned it to his undershirt. His big, lumpy face was sweet and gentle. "Thank you, friend," he said. Then, very softly, with his head hanging a little—"Ahm feelin' bad about this a-happening' t'me. Mah wife was a good woman." He paused. And then, as though talking to himself, so low I could hardly hear it, "Ahm feelin' right bad."

As he said this, I looked into his face. Slowly the life was going out of his eyes. It seemed to recede and go deep into the sockets like the flame of a candle going into the night. Over the eyeballs came that dull glaze. I had lost him. He sat deep within himself in his sorrowful, dark absorption.

That was all. We sat together. In me there was only mute emotion—pity and love for him, and a cold, deep hatred for what had killed him.

Presently he arose. He did not speak. Nor did I. I saw his thick, broad back in the blue work shirt as he stood by the door. Then he moved out into the darkness and rain.

CHAPTER FOUR

THROUGH THE EYES OF THE MEDIA: 1928–1936

From 1928 to 1936 local newspapers chronicled the planning, construction, trials and tribulations related to the Hawks Nest project. The majority of the articles in this chapter appeared in *The Fayette Tribune: The Home Paper with More Readers than all other Fayette County Papers*.[54]

As the newspaper accounts will show, Union Carbide originally planned to build two dams and two power houses; and in December, 1928, received an operating permit from the Public Service Commission for this dual construction. During the early part of 1929 company officials revised original plans, deciding instead to build only one dam, one tunnel and one power house. On July 25, 1929, the PSC[55] was approached with the revised plans—changes that also necessitated a revision of their original construction application. Newspaper articles published in the Fayette Tribune, July 31, 1929, outlined those revisions.

During the month of August 1929, railroad and tunnel contractors from around the southeast were afforded the opportunity to bid on the Hawks Nest project. According to a Fayette Tribune article on August 28, 1929, bids were opened in September of that year. It need be noted here that The New Kanawha Power Company accepted construction bids several months prior to receiving approval on their modified construction plan, which was not granted until December 1929.

Beginning in 1932, the history of the tunnel, as we know it today, began to emerge in the media. Articles related to silicosis-related law suits brought against Rinehart and Dennis and the New Kanawha Power Company dominated the news and culminated with silicosis trials in Fayetteville and Charleston from 1934 to 1935.

From the outset Rinehart and Dennis contended that compensation for the silicosis victims should be forthcoming from the state's compensation coffers; but unfortunately WV's compensation act did not cover silicosis. "In the spring of 1933, the legislature of West Virginia had considered a bill intended to compensate the tunnel workers afflicted with silicosis. It was defeated under strong pressure from industry."[56]

Author's Note: For ease of reading I have opted against the editorial convention of inserting "[sic]" to indicate odd or incorrect usage (including typographical errors) found in the original newspaper articles.
54. *Fayette Tribune* articles are located on microfilm at the Oak Hill Library, Oak Hill, WV.
55. Appendices: Request to "Modify Plans and Build Only One Dam and One Power Plant," pg. 214.
56. Cherniack. pg. 69.

In 1935 when the state's compensation statute was eventually upgraded, the inclusion of a statute of limitations clause virtually denied all hopes that victims might receive compensation. "The West Virginia statute provides that all claims must be presented within 1 year 'from the date of the last exposure to silicon dioxide dust in harmful quantities.' But what is worse, death claims must likewise be presented within 1 year. And as a further condition, a worker must have been employed for 2 years in the same employment."[57]

"If a worker has worked for 2 years in the same employment, if he presents his claim within 1 year after leaving his job, if he has given his life history in all detail to the employer, if he has never broken any safety rules, he may recover $500, perhaps $1,000, but no more though he should later be dying of silicosis."[58] Considering that the tunnel work at Hawks Nest lasted only eighteen months, this clause, for all practical purposes, insured that victims would not receive compensation.

As the newspaper articles dealing with the lawsuits will reveal, all of the silicosis cases were either dismissed or settled for a pittance. In fact, the first group of claims that initially totaled $4 million was settled for $130,000. Half of that settlement went to the attorneys and the remainder was divided among approximately 150 men. The proceeds were distributed according to the following compensation schedule established by Judge Eary: "$400 for an unmarried black man; $600 for a married black man; $800 for an unmarried white man; and $1,000 for a married white man. The families of deceased married white men should receive an additional $600, making $1,600 the maximum judgment."[59]

Compensation for the remaining plaintiffs not included in this initial settlement—those represented by the Charleston law firm of Townshend, Bock and Moore—was even less. Although their attorneys argued that the compensation statute of limitation clause was adopted after the cases were filed, but "seeing that further action had been rendered futile, Townshend, Bock and Moore acknowledged defeat and negotiated a block settlement with Rinehart and Dennis for seventy thousand dollars. The sum was supposedly divided among the remaining plaintiffs on 9 July 1935.[60] James Mason's statement to the congressional subcommittee covered the details of that settlement.

While the topic of "silicosis" occupied the center stage of the media's attention, Union Carbide and the State of West Virginia were concurrently fighting a battle of different sorts with the Federal Power Commission. As early as May 9, 1927, when the New Kanawha Power Company filed a notice of intention with the Federal Power Commission advising of their intention to dam the New River, they made it clear

---

57. "An Investigation. . . ." pg. 119.
58. Ibid.
59. Cherniack. pg. 67.
60. Ibid., pg. 72.

that, in their "corporate opinion," jurisdiction of the New River belonged to the West Virginia Public Service Commission. By challenging the FPC's jurisdiction as set forth in The Federal Power Act of 1920—an act which assigned oversight of all "navigable waterways and interstate commerce" to the FPC—the Union Carbide Corporation established the groundwork for a dispute that would last for over a quarter of a century. A correspondence dated January 23, 1928, from O.M. Jones and the WV PSC set forth the logic by which Union Carbide and the State of WV challenged the FPC's jurisdiction over the New River in Fayette County.[61]

The issue of jurisdiction was eventually tried in the U. S. Supreme Court in the trial of the UNITED STATES v. STATE OF WEST VIRGINIA, 295 U.S. 463 (1935)[62] No. 17; and while the Supreme Court's decision did not settle the question of jurisdiction, it removed federal pressure for several years. In fact Union Carbide operated the Hawks Nest Dam and Power House, without a federal license for more than three decades; and "…it would be February of 1967 before Union Carbide and the State of West Virginia acquiesced to the Federal Government's decision. On March 9, 1967, the Union Carbide Alloy plant was issued a Federal license to operate its hydro-facilities. The license was made retroactive to January 1, 1938, and was good for fifty years, expiring on December 31, 1987. On December 11, 1987, the Elkem Metals Company Alloy plant was issued a new license, good for thirty years, by the Federal Energy Regulatory Commission, successor to the FPC."[63]

1928

HYDRO-ELECTRIC POWER ON KANAWHA
Fayette Tribune, January, 1928

Every kilowatt of power produced by the New-Kanawha Power Company will be used to develop the industries along the Kanawha Valley, if it eventually is licensed to construct plants at Deepwater, and at the proposed new dam above Charleston, according to L.H. Davis, New York vice president of the company.

This statement was made by Mr. Davis after he appeared before Major A.D. Ardery, U.S. engineer, at the public hearing over applications for preliminary permits to construct such plants, filed by the New-Kanawha company and the West Virginia Power Company.

Mr. Davis declared that his company is vitally interested in the development of

---

61. Appendicies, "Memorandum of the WV PSC to the Federal Power Commission," pg. 212.
62. Appendicies, U.S. Supreme Court Ruling, pg. 223.
63. McKinney, Tim. *Elkern Metals*. Pictorial Histories Publishing Co., Inc. Charleston, WV. pg. 40.

the valley, pointing out that the parent company, the Union Carbide and Carbon Co., of New York already has vast operations in this section. He would not indicate what further development the company plans aside from the power projects and the proposed new metallurgical plant at Boncar which for the present is contingent upon the granting of the preliminary permit to his company.

Concerning the protest entered by the state at the hearing against the issuance of even a preliminary permit to either company, because neither has filed an application for a license with the public service commission, Mr. Davis expressed the opinion that it was a mere formality and that the state was acting merely to conserve its interests.

He explained that his company also had not filed an application with the public service commission because at this stage of the proceedings it is not technically necessary. He declared that his company would not fail to respect every right of the state involved and all requirements will be met in due time, he said.

Mr. Davis' statement, read before Major Ardery, describes in some detail, the projects contemplated by his company in the Kanawha valley at Deepwater and Locks 3 and 5. He does not describe, however the additional projects contemplated beyond Deepwater, plans for which provide for tunneling under the mountain at Hawks Nest and diverting the course of New River at that point.

This plan was described some time ago. It included construction of either one long tunnel, or two short ones. In the event a long one is constructed, one large power plant will be constructed at the lower end. If two tunnels are built, two power plants will be built.

However, it is understood that the tunnel projects are aside from the plans for Deepwater and Locks 3 and 5, and will be taken up by the company at a later date. The land on both sides of the river at the Hawks Nest location for a distance of 10 miles is owned by the company.[64]

Mr. Davis' statement follows: "New Kanawha Power Company in its application to the Federal Power Commission for preliminary permit for water power development on the Kanawha River, West Virginia, has presented a comprehensive project for the complete utilization of the entire fall of the Kanawha river from the foot of Kanawha Falls at Glen Ferris to the lower pool at the proposed new lock and dam No. 5 of the U.S. government near Burning Springs.

"The total fall in the stretch of river named is 63 feet and it is proposed to develop the power available from such a fall in three equal stages of 21 feet each with substantially identical power house and power equipment structures at each development. These three developments are to be located near Deepwater, one at the proposed new Lock 3 and one at the proposed new Lock 5. Such a plan of development will be conducive to the most economical cost, both of construction and of operation, and will result in the lowest practicable cost of power producible from the three power stations proposed.

---

64. Appendices. "Electro Metallurgical Company Hawks Nest Hydro Electric Development," pg. 194.

"The applicant proposes to install power generating equipment up to the economical limit at each of the three power sites. Its present view is that such limit will be represented by a capacity to discharge about 10,800 cu. feet per second at maximum efficiency, which will correspond to a total rated power capacity of about 22,000 horsepower at each site, under a head of 21 feet. The initial installation at each site will be for the use of at least 5400 cu. ft. per second, which represents the flow and therefore the power which will be available, on an average for about eight months or more during the year. The remaining installation will be put in as soon as the demand therefore requires it.

"The applicant proposes to utilize to such extent as it may find advantageous, the power from the proposed power stations for the expansion of the electro chemical and electro metallurgical operation of its associated companies, one of which has had for many years a hydro-electric development at Kanawha Falls, which is situated immediately at the head of the Deepwater development proposed in the above mentioned application. The products of these electro chemical and electro metallurgical commodities are among the most essential commodities used in the basic industries of this country both in peace and war.

"The applicant proposes to dispose of the remaining power for public utility use."

[last paragraph illegible]

HEARING ON PERMIT FOR KANAWHA-NEW RIVER POWER
Fayette Tribune, January 1928

Hearings on the applications for power permits of two[65] hydro-electric concerns who propose to use New river and Kanawha power will be held at the federal engineer's office at Huntington Thursday.

The Union Carbide Company, which is one of the applicants for permit, announces plans for development of its South Charleston and Blaine Island plant to cost 7 million dollars.

The Union Carbide and Chemical Corporation is said to have definitely determined several months ago to center on the Kanawha Valley as the location for all future plant extensions. It is said to own half a hundred chemical manufacturing plants at various places in the United States. In each of which millions of dollars are invested.

It is said that the company has almost unlimited resources.

The Electrometallurgical Company's plant at Glen Ferris, one of the largest plants in the Kanawha valley, is allied with this combination of capital.

Other plans of the company are said to be so immense as to almost dwarf the South

---

65. The New-Kanawha Power Company and Appalachian Power Company.

Charleston operation. The New-Kanawha Power Company, said to be a subsidiary of the Union company, announced last summer their plans in formation for the building of several great power dams and plants in the Kanawha and New Rivers, to cost about $23,000,000. There will probably be five of these dams within 35 miles of Charleston and each will have a large power house for the generation of electrical current that will be used in plants in their neighborhood.

It is said that a new industrial plant will be erected near Boncar[66] by the Electro-Metallurgical Company on a tract of approximately 600 acres purchased last summer. Applications for the construction of the dams have been filed with the United States government, and other dams and possibly a tunnel near Hawks Nest will be part of that project. It is said to be also the plans of the company to erect one of the dams at the head of navigation on the Kanawha and if cooperation may be had with the United States government, to include locks and extend navigation to Kanawha Falls.

It is estimated that the total electrical power that will be generated by the entire system will be around 100,000 horse power, which is twice as much as it takes to supply Charleston. All of it will be utilized for manufacturing purposes however, it is said. It is thought that the company will spend at least $20,000,000 on this system of dams and tunnels and that this amount will be augmented by an additional $3,000,000 by the federal government in extending the Kanawha river navigation facilities.

NOTICE OF AN APPLICATION TO CONSTRUCT A DAM
Fayette Tribune, July 4, 1928

TO ALL WHOM IT MAY CONCERN:

Notice is hereby given that the undersigned, a West Virginia public service corporation, will on the 31st day of July, 1928, at eleven o'clock A.M. apply to The Public Service Commission of West Virginia, at its office in the Capitol Building, at Charleston, for a permit authorizing the undersigned to construct a dam at a site on New River in Falls District and Ranch District, County of Fayette, West Virginia, about one-quarter of a mile upstream from the Blue Ash Tunnel of the Chesapeake and Ohio Railway Company, and in connection therewith a hydro-electric plant at a site on New River in the said Falls District about one-third of a mile upstream from the Gauley Junction Station of the Chesapeake and Ohio Railway company, and a tunnel connecting the said dam and the said hydro-electric plant, pursuant to and in accordance with the provisions of Chapter 54-B of the Code of West Virginia.

And notice is further given that the dam proposed to be constructed as aforesaid will

---

66. Appendices, pg. 194.

be approximately 47 feet high above low water and that the following described lands will need be required in the construction of said dam, said hydroelectric plant and said tunnel, or will likely be flooded by the water impounded from said dam . . . .

(The remainder of the notice dealt with property boundaries and descriptions.)

This the 20th day of June 1928

NEW KANAWHA POWER COMPANY
BY EDWARD S. WHITNEY, President[67]
Glen Ferris, West Virginia

PLANS OF NEW-KANAWHA CO. TO USE NEW RIVER POWER
Fayette Tribune, September 12, 1928

A large manufacturing plant at Boncar, one of the greatest industries of the Kanawha valley, to be operated by the Union Carbide and Carbon Company and an expenditure of nine millions for a New River hydro-electric power plant, are part of the plans of eastern capital proposed in connection with the utilization of New river power at Hawks Nest and Cotton Hill. Plans for the development were revealed by high officials of the New-Kanawha Power Co., at the hearing before the Public Service Commission.

O.M. Jones, the company's chief engineer, asserted that interference with the river's flow so as to affect navigation was extremely remote and could happen only in the event of an accident to one of the gates governing the flow of water through the tunnels which are to be constructed as a part of the company's project. Even then the interference would not exceed 12 to 18 hours duration, depending on where the accident occurred, he said.

The testimony was drawn out by Clyde B. Johnson who was attempting to show all of the facts concerning such a contingency as interference with navigation.

Only one protest against granting the permit was entered, coming from the Hatfield-Campbell's Creek Coal Company of Cincinnati,[68] which has a fleet of barges on the Kanawha river to transport its coal. Representatives of the Hatfield company were present and paid close attention to the testimony.

It was indicated that the coal company is interested only in obtaining assurances that there will be no interference of the normal flow of the Kanawha.

Mr. Jones said that his company did not intend to interfere with the rights of anyone

---

67. Edward S. Whitney would later serve as president of Union Carbide until 1956.
68. Appendices, Telegram, J.T. Hatfield to PSC, pg. 209.

else and every safeguard to prevent interferences of any kind had been taken in making the plans for carrying out the power project.

The purpose of the dams, Mr. Jones explained, is to divert the water through the tunnels at the mouths of which the water will flow into the turbines at the electric power plants to be constructed. The dams, he said, will be kept full at all times and 24 hours each day. *[next sentence illegible]*

J.J. Riley, of the Barium Reduction Company of South Charleston,[69] after saying his company had no connection whatever with the Carbide interests, declared that the proposed dams and power plants would prove of great benefit to the state. "They would be a great advantage to the Kanawha Valley and to the entire chemical industry of this section," he said. "It would be more advantageous to consume the power for manufacturing than for electric lighting purposes."

Chester C. Counts of the chief engineer's staff, C&O railroad, of Richmond, declared that his road has no objection to the construction of the dams and power plants, asserting that he has made a thorough inspection of the sites and declaring that they would in no way threaten the railroad.

At the conclusion of the hearing, the Public Service Commission announced that it had employed C.E. Krebs, geologist, and mining engineer, to investigate the project and report back his findings. Employment of an expert to make special studies for the commission is authorized by law. This is the first time the commission had had to consider a project of such magnitude under the present water power law.

The New-Kanawha Power company has been making plans for building two dams for a number of years and O.M. Jones, chief engineer, has spent most of his time during the past five years developing the plans. It was evident from the testimony of company officials that they had used every care to guard the rights and interests of the state as well as of individuals who might in any way be affected by the dams.

Leonard H. Davis, New York vice president of the company, told the commission that it was the plan of the company to consume virtually all of the power generated by the dam at a plant to be built at Boncar, where land has already been purchased. He did not go into details as to the nature of the plant but it is understood that it will be similar to the Electro-Metallurgical Company's plant at Glen Ferris.

[last sentence illegible]

---

69. Appendices, "The Chesapeake and Ohio Railway Company," pg. 210.

STATE MAY TAKE NEW RIVER DAM POWER AFTER 50 YEARS
Fayette Tribune, December 12, 1928

In accordance with the permission granted to the New-Kanawha Power Company to construct dams at Hawks Nest and Gauley Junction, two months ago by the Public Service Commission, a written order granting the petition and specifying the powers and limitation of the company in regard to the dams was entered by the commission Saturday.

Construction of the dams and the contiguous power house, which the company estimates will cost between $7,000,000 and $9,000,000, is to be commenced within one year from date of the acceptance of the permit granted by the Public Service Commission, according to the order.

The order sets out "that all rights under the permit shall terminate at the end of 50 years from the date of said grant, and that full control, occupancy and enjoyment of said permit shall at the expiration of said period of 50 years from the date thereof revert to and revest in the state of West Virginia, and the people thereof with full power and authority to make such disposition thereof as to the state shall then appear best.

"That if at the expiration of said permit, the same shall not be renewed and if the state shall elect, by itself or another agency to operate said power site, the state may at any time after said 50 years from the date of said permit, upon at least one year's notice thereof acquire all the property of the said New-Kanawha Power Company, its successors or assigns, acquired constructed or maintained by it or them and used and useful in carrying out the purposes for which said permit is granted and for which any permits, rights, franchises and privileges shall then have been granted by the state, or any political subdivision thereof.

"That in the valuation of the property to be acquired…no allowance shall be included for unreasonable costs of financing, for promoter's profits…or for the value of any franchise.

"Permit to build the dam is the first allowed by the commission under the Water Power Act of 1915."

1929

TWO ENGINEERS DROWNED IN HAWKS NEST RAPIDS
Fayette Tribune, May 29, 1929

Two young engineers in the employ of the New-Kanawha Power Co. were drowned in Hawks Nest rapids on New River Friday morning. A third man escaped by swimming to shore. The drowned men were Russel Richards of Columbus, Ohio and H. Armentrout of Glen Ferris. N. Perdue saved himself by swimming.

County Road at Hawks Nest Station, taken from Lovers Leap Rock, showing the construction track. *Courtesy West Virginia State Archives.*

The engineers were engaged in making soundings of the river bed in connection with the power development under way between Hawks Nest and Gauley Bridge. They were in a small skiff which was being lowered through the rapids by a steel cable held by men on the shore. The cable broke and the boat went into rough rapids and capsized, throwing the three occupants into the river. Parties on shore were unable to go to the rescue. All the men were good swimmers. Richards was within a few feet of shore when he sank. Armentrout was unable to get beyond the swift current and was carried down stream.

Perdue's experience and grief over the death of his companions has completely unnerved him and he is on the point of collapse.

The body of Richards was recovered within a few hours and that of Armentrout was found Tuesday morning.

Both men were married. Richards's wife's people live at St. Petersburg, Florida, where the body was sent for burial.

Three men have now been drowned while engaged in making river soundings incident to construction of the Hawks Nest dam and hydro electric development of the New-Kanawha Company. Waittman Teubert was drowned in a similar manner below Kanawha Falls last February. His body was recovered last week.

ONE TUNNEL ONE POWER PLANT, IS REVISED PLANS OF THE NEW-KANAWHA CO.[70]
Fayette Tribune, July 31, 1929

A tunnel under the mountain at Hawks Nest, three miles long, diversion of a portion of New River through it and the erection of a single power plant at the outlet constitute the revised plan of the New-Kanawha Power Company, made officially known by the filing of a petition with the State Public Service Commission, seeking authority to amend the company's original permit for the construction of two dams and two power plants.

As originally planned and authorized, the New-Kanawha Power Company was to construct two tunnels, 1 at Hawks Nest and the other in the vicinity of Gauley Bridge and power plants at the outlet of each tunnel. The Hawks Nest tunnel was to allow a fall of 80 feet and the Gauley tunnel was to allow a fall of 68 feet, making a total of 148 feet. In revising its plans the same total fall will be obtained from the one long tunnel and the single power plant will produce as much as two small plants, it is said.

In making the change, Mr. Davis, vice president of the New-Kanawha Company of New York, told the commission that 15- foot crest gates would be added to the height of the proposed dam at Hawks Nest, thereby making flood control more satisfactory.

The revised plans also eliminate the necessity of construction two dams as under the original plan two were contemplated.

Flood dangers in the river would be lessened, it was pointed out to the commission by O.M. Jones, chief engineer, who also was examined by Mr. Johnson by the elimination of the Gauley tunnel and dam, which had contemplated a pool of 810,000,000 gallons of water.

As for economy, the commission was told that the extra cost of the longer tunnel and higher crest gates would be more than offset by the savings affected by abandoning the second dam.

No date has been set for receiving bids on the tunnel, but some of the largest contractors in the country are interested in the project and have been examining engineer's plans. A time limit has been set for completion of the contract which would require work to be rushed with all possible speed.

It is unofficially reported that the stone from the tunnel is to be shipped to the site of the proposed new manufacturing plant at Boncar. To transport the stone from the tunnel workings will require the building of about six miles of standard gauge railroad

---

70. Appendices. "Request to Modify Plans and Build Only One Dam and One Power Plant," pg. 214, and "Letter from Campbell's Creek Coal Co.," pg. 214.

track. This will be an expensive undertaking in itself, as the roadbed will have to be blasted along the high canyon walls for a considerable distance.

It is also reported that tunnel operations will be carried on at both ends and a third point near the middle giving working space for four crews. The New Kanawha Power Company is a subsidiary of the Union Carbide and Carbon Company of New York, which operates plants at South Charleston, Glen Ferris and other points. The power to be generated at the proposed new plant will be used in manufacturing plants to be erected by the company, the principal plant being scheduled for erection at Boncar on land already owned by the company.

NEW KANAWHA COMPANY TO OPEN BIDS SEPTEMBER 3RD ON TUNNEL AND DAM
Fayette Tribune, August 28, 1929[71]

Bids will be opened September 3 by the New Kanawha Power Company for the construction of tunnels and other improvements to be made below Hawks Nest for the purpose of developing New River hydroelectric power.

The project is one of the largest under way in the country, involving an expenditure of over 10 million dollars. But few contractors of national reputation are in position to consider the giant undertaking. Four of these contracting firms have had engineers going over plans and construction details for several months. The two big items in the contract are the building of a dam in New River and boring a 3 mile tunnel through the mountain to carry diverted current of the river to the turbine plant near Gauley Bridge.

It is understood that the millions of yards of stone from the tunnel workings will be shipped to Blaine Island in the Kanawha below Charleston and used to rip-rap around the island. The stone would probably be shipped by rail to Boncar and there loaded onto barges and floated to the island.

The time estimated to complete the job is about 3 years. About 800 men would be employed on the work. Labor camps will probably be established at Hawks Nest and on the old Tompkins farm owned by the Electro Metallurgical Company, a subsidiary of the Union Carbide and Carbon Company, which is the chief power behind the whole enterprise.

---

71. Appendices. "Examples of Bid Items," pg. 204.

1930

FORTY ENGINEERS WORKING ON HAWKS NEST PROJECT
Fayette Tribune, January 29, 1930

Speaking to the Charleston chamber of commerce, Owen M. Jones, of Glen Ferris, chief engineer of the New-Kanawha Power Co., told of the Hawks Nest development of his company, which includes the digging of a tunnel three miles long to divert the New River and construction of a power plant, declared that 40 men are at work perfecting the technical details of the project. He added that his company is not overlooking the value of coal as a power producer. This was taken to mean that a coal burning power plant may be constructed to operate in conjunction with the water power plant, already authorized.

CONTRACTOR SIGNING UP FOR HAWKS NEST POWER PROJECT
Fayette Tribune, February 19, 1930

While no official announcement has been made, it is learned on reliable authority that the contract for the Hawks Nest hydro-electric project of the New-Kanawha Power Company is to be awarded to Rhinehart, Dennis Company of Charlottesville. A representative of that firm was called in New York last week to sign up the final papers. Rhinehart Dennis Co. is one of the oldest and largest railroad contracting firms in the country. The contract for tunnel, dam and railroad grading will amount to about two and a quarter million dollars.

CONTRACT LET FOR HAWKS NEST DAM AND TUNNEL[72]
Fayette Tribune, March 19, 1930

The contract for construction of the three mile tunnel and dam at Hawks Nest was awarded by the New-Kanawha Power Company, owners of the project at New York, Friday to the firm of Rinehart and Dennis of Charlottesville, Virginia.

It was understood several weeks ago that the Charlottesville firm was low bidder on

---

72. Appendices. "Provisions of Contract between Rinehart and Dennis...." pg. 203.

the project. Rumors that no contract would be let for the present were unfounded as plans for the development of the new plants have gone forward as fast as practicable considering the enormous undertaking.

The figures in the contract were not announced, but estimates given to the State Public Service Commission in applying for a permit placed the cost between $8,000,000 and $9,000,000.

P.H. Faulconer, president of the firm of Rinehart and Dennis, of Charlottesville, VA spent Monday and Tuesday at Glen Ferris. They were accompanied by L.H. Davis, vice president and O.M. Jones, chief engineer of the New Kanawha Company. The bid was said to be about $4,000,000.

Work of building the dam and tunnel will start April 1 and will be completed in about two years, according to estimates.

The dam will be erected at a point in New River just below Hawks Nest and its purpose will be to divert the water through the tunnel which is to measure 16,250 feet. At one end of the tunnel a power station will be erected and the cost of this project contract for which will be awarded later, in a separate contract, will be about $750,000.

The capacity of the power plant will be 30,000 to 40,000 horsepower. The primary power production will be utilized by the company at Boncar, where a large metallurgical plant is being built by the Union Carbide interests.

The New Kanawha Company is a subsidiary of the Union Carbide and Carbon Company of New York, one of the largest holding concerns in the country, operating as it does, numerous manufacturing plants under different names, include the Carbide & Carbon Chemicals corporation at So. Charleston and the Electro-Metallurgical Company at Glen Ferris.

The plant under construction at Boncar, it is said, will be several times as large as the plant at Glen Ferris, where about 150 men are employed. To operate the new plant, an immediate amount of electric power will be required and it is to come from the new power project planned for Hawks Nest.

[Remaining article illegible.]

HAWKS NEST DAM TO BE DONE IN 2 YRS: COST $4,225,737
Fayette Tribune, March 26, 1930

The Public Service Commission on Wednesday referred the contract between the New Kanawha Power Company and Rhinehart and Dennis, of Charlottesville, Va., for the construction of a dam, a power house and a tunnel at Hawks Nest to Charles E. Krebs, engineer of Charleston for a report. The engineer, it was said, at the commission offices will report in the next month.

The contract, as submitted to the commission, calls for the start of the work April 1, to be completed by March 13, 1932, two years after the award is made.

The contract price of $4,225,737.50 provides for, among other things, the construction of the dam exclusive of the crest gates; of the intake; of the power house, the tunnel between the power house and dam; of a railway spur, the roadway to the intake and of the transmission lines.

Officials of the engineering company inspected the site for the development.

GIANT UNDERTAKING UNDER HEADWAY ON NEW RIVER
Fayette Tribune, April 2, 1930

Informal but enthusiastic was the observance of the beginning of actual construction work on the $9,000000 hydro-electric power development of the New-Kanawha Power Co., between Hawks Nest and Gauley. The event was celebrated Sunday when about 200 people gathered at Gauley Junction and witnessed the first moving of dirt on the big project. The affair was by way of honoring O.M. Jones, chief engineer who has been engaged on the engineering problems in connection with the undertaking for the past seven years.

Dr. W.P. Simmons, high official of the power company, presided at the ceremonies. Rev. Wm. Musick, of Gauley Bridge, led a prayer for the success of the project, and Geo. T. Lancaster of the Electro-Metallurgical Co. of Glen Ferris, a subsidiary of the Union Carbide and Carbon Corporation, delivered the opening address.

After describing the dream of 35 years which is culminating in this project, he paid a tribute to Mr. Jones and particularly referred to the vision and foresight of Major Moorehead, founder of the carbide interests.

T.R. Ragland, of Beckley, original superintendent of the Glen Ferris plant and an associate of Major Moorehead, recalled to the minds of the assembled crowd the activities of the past and joined in the tribute to the engineer of the present.

I. Wade Coffman, chairman of the State Public Service Commission, said that West Virginia looks with great favor upon the project and Clyde B. Johnson, attorney for the power company, described in detail the enormous industrial and commercial benefits which he believes will result from this and other enterprises to follow:

For a few minutes the celebration assumed the likeness of a ship launching as the great steam shovel was christened.

Mr. Jones was asked to speak, but characteristically he set aside the opportunity to talk when there was a chance for action at hand. He climbed to the platform of the steam shovel and sent the great prongs into the earth to turn out the first shovelfull

Gathering to honor Mr. Jones and the starting of construction. *Courtesy West Virginia State Archives.*

of dirt for the railroad which will run into the power plant. A buffet supper was later served by the power companies employees at the Glen Ferris Inn.

Contract for the dam at New River at Hawks Nest and the three mile tunnel under Gauley mountain has been let to Rinehart and Dennis, Charlottesville contractors, for $4,225,737. Terms provide for completion by March 13, 1932.

The first task of the contractor will be the construction of about 10 miles of railway along the north side of New River from Hawks Nest to Gauley Junction. Work will go forward at several points as soon as steam shovels can be placed. Shovels began operation at Gauley and Hawks Nest Monday morning. Clearing of the right of way will give employment to a large force at once. There is a plentiful supply of labor, the contractor paying 30 cents an hour for 8 hour shifts.

Many buildings are being constructed at Gauley and Hawks Nest to serve for housing, labor storage and work shops. A fully equipped hospital will be maintained by the contractor.[73] Several hundred men will be employed on the work for two years. Every

---

73. In fact, the four main worksites only provided first aid kits. Two company doctors served on an "as needed basis" to handle minor accidents. Injuries of significance were transported to the Coal Valley Hospital in Montgomery.

minute detail of the undertaking has been worked out in advance by the engineering department under the direction of Chief Jones.

[last paragraph omitted]

AWARD $1,000,000 CONTRACT FOR BIG TURBINE GENERATORS
Fayette Tribune, September 24, 1930

The New-Kanawha Power Company, which is building a hydroelectric power project at Hawks Nest and Boncar, including the big 32 foot three mile tunnel under the mountains, and a dam in New river, has let a contract to the Westinghouse Electric and Manufacturing company of Pittsburgh, for the installation of generators at the new power project. The total amount involved is approximately $1,000,000. The contract calls for delivery and erection of four vertical water wheel generators with direct connection and pilot exciters.

Construction of the tunnel and dam started early the past spring, and during the summer considerable progress has been made. The total cost of this project will be around $9,000,000 and completion is expected in about 18 months.

FIRST UNIT BONCAR PLANT TO BE STARTED
Fayette Tribune, October 8, 1930

Contracts for the first unit of the new plant of the Electro-Metallurgical company to be built at Boncar will be let within about ten days. This project will cost approximately $250,000.

Bids on the several buildings which will constitute the new plant were to arrive in the main offices of the company at Niagara Falls, N.Y., Monday and will be opened sometime the last of this week. The awarding of contract will be made within about 10 days. The second unit of the plant will not be started until next summer.

Power to operate the new plant will be supplied from the New-Kanawha Power Company now going up in the vicinity of Hawks Nest on New river. This company has a plant in operation at Glen Ferris and obtains power from the Kanawha Falls dam. Both companies are subsidiaries of the Union Carbide & Carbon Co., of New York.

Materials to be used in the construction of the unit will be of steel, corrugated iron and concrete. A motor house, locomotive house, track scale, yard office and a furnace and packing house are to be built. Buildings alone will cover almost 10,000 square feet of space.

Transformers during construction. 20,000 KVA single-phase transformer being untanked. *Courtesy West Virginia State Archives.*

Both the new power project and the plant at Boncar are expected to be in operation within a little over a year.

GLEN FERRIS ENGINEER IS GIVEN APPOINTMENT TO BOARD
Fayette Tribune, December 10, 1930

Owen M. Jones, chief engineer for the New Kanawha Power company of Glen Ferris, was appointed to the state board of registration for engineers by Governor W. G. Conley. Mr. Jones was appointed for the term ending June 30, 1934, and succeeds Carl L. Horner, of Clarksburg.

1931

CONTROL EQUIPMENT ORDERED FOR HAWKS NEST PLANT
Fayette Tribune, February 4, 1931

A $250,000 contract for equipment was placed by the New Kanawha Power company, subsidiary of the Union Carbide and Carbon company, with the Westinghouse Electric and Manufacturing company at Pittsburgh last week.

The contract calls for delivery of complete equipment, including an eight panel switch-board, a 10 panel auxiliary board and 69 KV high tension equipment for control of four 30,000 KVA waterwheel generators previously ordered.

This equipment will be used for control of the generators and the switching to the transmission lines. As yet no announcement has been made of an order for the transformers and transmission equipment, other than the oil circuit breakers referred to above.

CHARLESTON FIRM GIVEN BONCAR BRICK CONTRACT
Fayette Tribune, February 18, 1931

Contract for brick work at a plant being erected at Boncar by the Electro-Metallurgical company has been awarded to H.G. Agsten and company, of Charleston, for $38,000 it was announced Monday.

A mix house, 840 feet long, is now being built at the plant at a cost of approximately

![Power House Control Room]

Power House Control Room showing the main and auxiliary switch boards. *Courtesy West Virginia State Archives.*

$350,000. It will contain a trestle 600 feet long for use in carrying railroad cars from the second floor to the mix house, and is but one unit of the plant.

PROPOSAL MADE TO CHANGE NAME BONCAR IN PETITION TO COURT
Fayette Tribune, March 11, 1931

County Court Adjourned Tuesday Until March 13 by Pres. Tully.

A proposal to change the name of the town of Boncar, West Virginia, to "Alloy," West Virginia, was contained in a petition on behalf of the citizens of Boncar which was presented Tuesday before the Fayette county court by James C. Conley, of Charleston.

A quorum of the court was not present Tuesday and adjournment was made until Friday, March 13, by President L.S. Tully, who was the only member of the court

Dam during construction. Cofferdam being extended to include blocks 20 and 21. *Courtesy West Virginia State Archives.*

present. It is understood that the petition will be taken under advisement at this time. The county court as well as the post office department must act favorably upon the petition to make it legal, it was stated.

[last paragraph omitted]

FIND NEGRO IN RIVER
Fayette Tribune, May 13, 1931

The dead body of an unidentified Negro was found Tuesday in New River in the cofferdams above dam No. 3, by workmen for Rinehart & Dennis. It was stated that the body evidently had been in the river for two or three weeks. There were no marks on the body that would indicate foul play, it was said.

CHIEF R.M. LAMBIE MAKES INVESTIGATION AT TUNNEL
Fayette Tribune, May 20, 1931

Investigation of working conditions and loss of life among workers employed by Rinehart and Dennis on the tunnel project at the New Kanawha Power company on the New River began Tuesday when Robert M. Lambie, chief of the state department of mines came here from Charleston, accompanied by a group of men from this department.

Chief Lambie was accompanied on his inspection by Sheriff C.A. Conley and some of his deputies and Prosecuting Attorney H.E. Dillon, Jr. The chief left behind him two associates who will remain on the job for several weeks.

The investigation at this time was precipitated by an unusual number of deaths last week through accidents and disease. This has been high especially among colored workers. Among the six deaths known to have taken place last week was that of Howard Shepherd, of Gauley Bridge, who was killed in an accident.

Rumors and reports of various conditions known to exist at the tunnel and dam project are discussed generally by residents but little first hand information is obtainable because of the "gag rule" enforced by executives of the contracting company which employs about 1200 men. Officers of the law state that company officials refuse to converse with them and laborers do so with the fear of losing their jobs.

The *Tribune* has been appealed to on more than one occasion to lay bare these atrocious conditions against humanity, but facts obtainable have not been provable. The *Tribune* hopes to have Mr. Lambie's report by next week for publication.

Officers say that officials of the New Kanawha Power company are not responsible for the conditions, but lay the blame to the contractor.

ARMY OF WORKMEN DRILLING THROUGH GAULEY MOUNTAINS
Fayette Tribune, June 3, 1931

Big Power Project of Union Carbide Company Results in Unexpected Discovery of New Source of Wealth
(From the *Charleston Gazette*)

Steadily forging ahead at the rate of 20 feet a day or more, the big force at work with explosives, steam shovels and other mechanical devices are slowly eating a hole through the Gauley mountains.

Taken from the Midland Trail of the surge basin location and spillway. *Courtesy West Virginia State Archives.*

A huge tunnel, 32 feet in diameter, three and one-eighth miles long is being bored through which New river which has followed for forgotten centuries the old course, will hereafter flow in order that man can harness for electrical purposes the power contained in the rushing waters. And to make doubly sure that the waters that will turn the giant turbines will have their maximum power, the tunnel has a steady fall of 168 feet from one end to the other.

Until the Union Carbide company launched its $16,000,000 project into the bowels of the earth below Hawks Nest, the New river canyon was chiefly famous for its scenic grandeur. It was one of this state's main "selling points" to lure tourists into the West Virginia hills. No one could gaze down into the gorge and not come away without a lasting impression of the rugged mountains rising all about, beautiful in their bleak ugliness.

But modern man came with his passion for mastering the works of nature, and seeing that the hard and barren rocks, gashed everywhere with water courses, set about to put a bridle on the roaring, plunging wild river. The task is slow and tedious because of the ruggedness of the country, but today there is taking form one of the greatest engineering conquests of the present era.

Time is an ever important element when corporations the size of the Union Carbide company sets out to finish a job. So instead of starting at one end of the tunnel and driving through to the other, the tunnel is being driven in four directions at the same time. About the middle of the three mile stretch, between Cotton Hill and Gauley, nature has placed a convenient ravine. Driving in to the dead center of where the tunnel is to go, crews set out in opposite directions to meet other crews coming in from without. And the marvel of it all is that when four crews meet and the last partition of stone is hoisted out, the tunnel will be resting where the engineers intended it to rest, right on dead center.

Work on the lower end, (Gauley) of the tunnel has progressed to more than 4000 feet on the upgrade to within 700 feet of the tunnel coming downgrade from center, while the upper end, (Hawks Nest) has been bored down toward the tunnel going up to meet it from the middle ravine, to within a distance of about 2200 feet. Contractors have marked on the calendar a date some six weeks hence when the hole will be bored from the lower end to the center. Already plans are being made for a celebration to fittingly observe the event.

There are between 1000 and 1200 men employed on the project. They work in two shifts of 10 hours, each day. There is a two hour resting period between the two shifts in order to permit the tunnel to be cleared, as well as it can, of the harmful gasses and smoke caused by the explosives. The rock is drilled and packed with powder during the early hours of the shift and at quitting time the blasts are set off. Then there is a two hour wait before the next crew goes in to let the smoke and gases drift out.

Boring of the rock is being done on what is technically called two "benches" which might be more easily understood if one can vision two giant steps. The crew on the upper "bench" is always about 10 feet farther into the tunnel than the lower crew, so as to make the handling of the loose rock easier.

The tunnel is now semi-circle in shape, but when it is finally dug through, there will be placed inside a tube of thick concrete to reinforce it against the water resistance. At the upper end, the river is being dammed to divert its course into the tunnel instead of around its natural course in which it has been rushing for untold centuries. The distance the river takes at present from the mouth of the tunnel to its outlet is about five and one-half miles, and the tunnel's length slightly more than three miles, gives a picture of the direct line the tunnel is taking.

As a precautionary measure provisions are being made for two giant reservoirs to care for the overflow from the tunnel in the event it is necessary to close the gates below, part of the water will be thrown into the lower reservoir and the rest in the middle. The middle reservoir will be conveniently located to the present course of the river, so the overflow can be diverted back into the old river bed.

The power house will generate electricity for the company's plants which will manufacture an alloy for steel. This alloy is known as "ferro-chrome" and will constitute the highest test steel known and will be used with other steel in a hardening process.

Part of the company's unit now located at Niagara Falls will be moved to the West Virginia plant, and the indications are that the company's operation when finally completed and under way will be larger than its operation at the Carbide and Carbon plant in South Charleston.

Like a tale from the story of Aladdin's lamp, boring of the tunnel has enriched the Union Carbide company with untold wealth. In the process of removing the rock, the workers came across a vast deposit of silica sandstone which assays 99.44 percent pure. It is as fine a grade of sandstone and especially adaptable for steel and glass work as has been found in the world. Discovery of this sandstone in the lower end of the tunnel brought about a big change in the operations for the excavations were immediately extended in size and the tunnel considerably enlarged.

The beginning of mining of this silica sandstone gave the state official supervision in the operation because the law creating the department of mines specifically puts sand mines under the supervision of the state mining department, and Chief R.M. Lambie has placed field men of the department on the operation and the contracting company is cooperating with the state regarding laws governing safety measures to be observed.

Plans at present call for the completion of the tunnel by the fall of 1932, and in the meantime work is going along on the construction of the plant at Boncar and already a good sized town has grown up in that vicinity.

CEREMONY TODAY TO MARK COMPLETION OF PART TUNNEL
Fayette Tribune, August 12, 1931

Invitations have been issued by the Rinehart & Dennis Construction company, contractors on the tunnel job for the New Kanawha Power company on New River for the celebration of the completion of the work of the tunnel.

The ceremony will be held Wednesday (today) from 4 p.m. to midnight, according to the notices. Following the celebration a barbecue dinner will be served at six o'clock and dancing will continue until midnight at Lovers Leap.

Work on the tunnel and dam began more than a year ago. The power plant to be constructed at the lower end of the tunnel, just above Gauley Bridge, will be a part of the development which is to provide power for the plant being constructed by the Union Carbide interests at Alloy.

The celebration is for completion of boring of part of the tunnel but it is yet to be lined with concrete which is still a large task.

GOVERNOR CONLEY ATTENDS TUNNEL CELEBRATION
Fayette Tribune, August 19, 1931

West Virginians, both prominent in private and public life, joined Wednesday afternoon with the personnel of power and contracting companies organizations in a celebration of the practical completion of the water diversion tunnel between Hawks Nest and Gauley.

More than a thousand invited guests gathered at the commodious club house at Lovers Leap to hear and participate in the program which continued from four o'clock in the afternoon until midnight.

Governor Wm. G. Conley, I. Wade Coffman, chairman of the public service commission; G.H. Davis, vice president of the New-Kanawha Power Company, and Hollis Dennis of the contracting firm which is installing the hydro-electric power development immediately above the junction of the New and Gauley rivers, made addresses. Other features of the program were music by the Gill and Schadel orchestras and dancing by the Haviland sisters of Charleston.

A dinner served following the entertainment features, was novel to the majority of the guests. The Lovers Leap club house is a large stone building of Old English design. It occupies a site on the crest of a mountain overlooking New River gorge, more than one thousand feet below. At this point the dam of the New-Kanawha Power Company which will divert the waters of New River through a 30 foot tunnel to a point three and one quarter miles below will be built. There, at the lower end of the tunnel the celebration was held.

Amid the wild beauty of the New river mountains under the canopy of arching forest trees two tables, 100 feet in length had been set off for the day's visitors. At six o'clock the guests were invited to dinner. As they approached the tables they passed the scene of the barbecue where over a trench six feet deep and 100 feet long, with smouldering fires in its depths animals were being roasted or barbecued whole. Mutton, pork and roast ox constituted the principle part of the menu, garnished with many appetizing accessories.

Beginning at eight o'clock there was dancing at the club house and this continued until midnight.

Big Project Costing Nine Million[74]

Facts relating to the history and objects of this monumental construction project were told the assembled hearers by the various speakers. It was stated that the dam, tunnel and power houses and incidentals expenditure will total $9,000,000. The dam across New River at Hawks Nest will divert the larger part of the streams' water through the tunnel which is 16250 feet in length. The lower, or western end of the tunnel is

---

74. Appendices. "Statement of Costs," pg. 205.

a short distance east of Gauley Bridge on the Midland trail. About here, an immense power plant having a capacity of developing 100,000 horsepower in electricity will be constructed. The power will be used in developing the natural resources in the southern section of West Virginia. It will be available for commercial use at low rates to consumers. The engineering feat of boring the two sections of the tunnel which on August 6, were "broken through into each other within an inch of the calculated direction," is considered a remarkable engineering feat. The completed section of the tunnel is 15,368 feet long. The borings were started at opposite ends. It was the job of the engineers in charge to make such accurate calculations and measurements that the tunnel sections would meet in both grade and on one center line. An added difficulty was a curve in one tunnel section; another was that the ground under which the tunnel runs is so rough that it was found impossible to measure a line directly above the borings. O.M. Jones, chief engineer for the New-Kanawha Power Company, R.E. Buckley, construction engineer, and P.J. Welsone, designing engineer, were given credit in some of the speeches for surmounting this difficult obstacle. Governor Conley stated that the successful completion of the tunnel section is a greater engineering feat than the boring of the water supply tunnel under the Hudson river in New York city.

Highly Commended

Clifford Waugh, J. Gooch and John Bolton, men who drove the work through in record time, day and night, under the supervision of Robert Perkins, Harry Faulconer and Linwood Faulconer, division superintendents, were highly commented by company officials.

The total number of cubic yards removed from the tunnel to date is given as 507,336. One day and night of a tunnel crew "drove" one of the heading 120 feet in six days. This was mentioned as a remarkable feat. This is a demonstration of "the real men" who have "helped put this job through" as one of the speakers remarked. The first tunnel heading was begun June 13, 1930, and the second September 25, 1930.

Among those present on the speaker's rostrom were Governor Conley, former Congressman J. Alfred Taylor, State Senator Clyde B. Johnson; G.H. Davis, vice president of the New-Kanawha Power company; Hollis Rinehart, of the contracting firm of Rinehart and Dennis, of Charlottesville, Va; T.H. Faulconer, president of the same company; and E.J. Perkins, vice president; O.M. Jones, chief engineer and P.J. Welsone, designing engineer; A. Stokes, an attorney whose residence is in Virginia made the address of welcome for the contracting company. His talk was well received. O.M. Jones was toastmaster.

Work on the construction projects is being pushed. Practical completion is expected within another year.

1932

ANOTHER ACTION AGAINST RINEHART AND DENNIS CO.
Fayette Tribune, August 10, 1932

Lee Wyatt has instituted suit for $25,000 damages against Rinehart & Dennis Construction company for personal injuries. The institution of this action brings the amount of damages up to $305,000 sought from the company by various plaintiffs who have instituted suits within the last few weeks. Love & Love are attorneys for Lee Wyatt.

NEGRO TUNNEL WORKER FOUND DEAD AT CAMP NO. 2
Fayette Tribune, September 7, 1932

Parker Brown, a Negro tunnel worker at Camp No. 2 on the big Hawks Nest power project, was found dead Thursday, September 1. The sheriff's office was notified by Doc. Russell, and Deputy Sheriff Ben Scarbrough investigated. Brown was found in one of the houses at Camp No. 2. The body was brought to the Dodd Undertaking establishment Thursday night. Death was probably due to some kind of gas.

Fayette county authorities are trying to get into communication with his sister at Knoxville, Tenn. The Knoxville chief of police is assisting in trying to locate her. Parker was originally from Atlanta, GA.

EIGHT MORE SUITS FILED
Fayette Tribune, September 28, 1932

Eight more suits were instituted in circuit court against Rinehart & Dennis company. Each of these suits are for $25,000 and this now brings the total to $785,000 in damages sought of this contracting firm.

The following were plaintiffs in these suits: Andrew Williams, Jno. Drain, Ernest Lyes, Dewey Wade, Aaron Stevenson, B.F. Johnson, Arlie B. Vanetter and Thomas C. Sheppard. Attorneys W.E. Teubert, L. Burke O'Neal and Love & Love represent the plaintiffs in these suits.

PLAINTIFF IN SUIT DIES
Fayette Tribune, September 28, 1932

Cecil L. Jones of Vanetta, former employee of the Rinehart and Dennis Construction company, on the New-Kanawha Power project, died Sunday morning at his home. It is said he is the second of the Jones family to die from silicosis. The deceased had instituted suit in circuit court here against the construction company. The action now will probably be in charge of his administrator.

TEST CASE IN DAMAGE SUITS BEFORE THE SUPREME COURT
Fayette Tribune, November 23, 1932

The state supreme court on Tuesday, docketed the case of Dora Jones, administratrix of the estate of Cecil L. Jones, against Rinehart & Dennis Company, which was certified from the circuit court of Fayette county.

The suit involves claims for damage for death resulting from his vocational occupation, and the supreme court will decide whether the administratrix was entitled to recover from the compensation department or sue for damages.

Jones, it is alleged, died as a result of silicosis, a disease contracted while in the employ of the contracting firm on the tunnel and the dam construction at Hawks Nest in this county.

When the case came before the circuit court, the judge overruled the defendant's demurrer to a declaration and certified the case to the supreme court.

The plaintiff originally sued for $5000 but on the death of Jones the amount was changed to $10,000 the limit provided in West Virginia law for death.

The decision in this case will have an important bearing on a number of similar suits pending against this same company.

VANETTA MAN ANOTHER VICTIM OF SILICOSIS, DEC. 7.
Fayette Tribune, Dec 14, 1932

Charles Means, 24 years old, died last Wednesday at his home at Vanetta of silicosis, a disease peculiar to workers in silica rock. Means was a former employe of Rinehart

& Dennis Construction company, contractors of the tunnel and dam project on New river at Hawks Nest. He was one of about 60 men who have filed suit against the company for damages resulting from work in the tunnel in which it is claimed the company was negligent in providing safe working conditions.

Already several deaths have resulted among workers who had been employed. Analyses performed on one or two preparatory to institution of the suits disclosed physicians said, that the deaths were caused by silicosis. There is no means of combating the disease once it has gained sufficient headway to be defeated, and it has been held that additiona deaths among workers are inevitable.

Funeral services for Means were held from the Summerlee church at 10 o'clock Friday morning. He was buried in the Bays cemetery.

Surviving him are his parents, Mr. and Mr. James Means of Vanetta.

1933

200 COAL MEN INSPECT THE NEW KANAWHA CO.
Fayette Tribune, January 11, 1933

A colossal engineering project which will ultimately divert New River from its present channel at Hawks Nest to flow under a mountain, was inspected last Saturday by more than 200 guests of the New-Kanawha Power company, in a tour of the tunnel at Hawks nest now rapidly nearing completion under the supervision of Rinehart and Dennis company.

The tour of inspection was made possible through the courtesy of Owen M. Jones, chief engineer of the New-Kanawha Power company. Included in the party were members of the New River & Winding Gulf Mining institute, headed by district mine inspector, Robert Lilly.

Congregating at the power house end of the 16,252 foot tunnel, the coal men first viewed the partially completed power house—a magnificent brick building, housing generating equipment—before boarding flat cars to be pulled virtually the entire distance of the tunnel itself, a journey that lasted for 42 minutes altogether. Afterward the most of the party went over the highway to the tunnel intake near Hawks Nest, to view the dam now under construction there to force New river into the tunnel portals for the underground shortcut.

Plant Is Explained

As explained to visitors by E.J. Perkins vice president of Rinehart and Dennis, and manager on the tunnel project, the plan to develop power by diverting the river from its course sounded very simple at first.

By means of a dam across the river at the intake end of the tunnel, the stream will

Power house during construction, showing steelwork completed. *Courtesy West Virginia State Archives.*

be forced to travel three miles by a direct route to the power house instead of six and one-half miles around the old river bed and on an indirect and unconfined route.

The effect of the tunnel is the same as would be obtained if New river were poured into a perpendicular shaft 162 in length with turbines at the lower end of the shaft, save for a slight difference in velocity due to the greater friction imposed upon the stream by the tunnel.

The tunnel, in other words, gives a 162 foot head at the turbine wheels in the power house.

Trip Through Long Tunnel

The train of flat cars bearing the visitors passed under an arched entrance similar to the opening of any railroad tunnel, but thereafter the similarity became less and less. Far ahead glimmered a seemingly endless row of incandescent lights, each casting a radiance that, beating upon the walls of the man-made hole, created the impression of hundreds of rifles twining about the interior of a monster gun barrel.

Photographer at penstock junctions. *Courtesy West Virginia State Archives.*

Tunnel Dimensions

The diameter of the tunnel, Mr. Perkins said, varies from 31 feet to 46 feet. For a total distance of 10,200 feet the tunnel is lined with concrete. The remainder of the bore is unlined, the solid rock through which it was driven being sufficient in texture to confine the water. Of this unlined portion 1000 feet is 42 feet in diameter and the remaining 5052 foot segment is 46 feet in diameter.

The variation in diameter, Mr. Perkins explained, compensates for variations in stream velocity in the various sections of the tunnel. In the unlined portions, the rough stone surface heightens friction while in the 10,200 foot concrete lined area velocity is greatly increased by the perfectly smooth walls.

The visitors gained some ideas when they observed a power shovel, parked to one side of the miniature railway which has been laid down through the bore for purposes of construction, with plenty of space left in which the train could pass.

Costs Nine Millions

When the project, estimated by Perkins to cost between nine and ten million dollars,

Transmission line construction. View of completed towers Nos. 22 and No. 23. *Courtesy West Virginia State Archives.*

is put into operation some time after September 1 of this year, the mighty force of New river, intensified by confinement, will be taken at the tunnel mouth, split into four separate streams and fed under a colossal pressure into four turbines, each of them attached to a generator that will produce 30,000 horsepower.

The steel shaft connecting turbine to the generator is 30 inches in diameter, 30 feet in length. Each turbine is suspended from a bearing capable of withstanding a 200-ton strain.

Power from the project will be taken over a line now under construction a distance of eight miles to Alloy, there to be used entirely in the manufacturing enterprises of the Union Carbide Co., of which the New Kanawha Power company is subsidiary. It will be used in the manufacture of alloys, ferro silica and manganese.

The chronology of construction is brief. In March 1930 work was begun by contractors—erection of camps, etc. On June 15 the business of punching a hole through solid rock—making the tunnel—was started. More than a year later, on August 7, 1931, crews holed the first heading, meaning that a crew which started in the middle of the tunnel "met" a crew which started at the Hawks Nest end. And slightly more than a month later, September 19, 1931, crews met on the other end of the tunnel. Since then construction of the power house, dam at the Hawks Nest end, and cementing of the tunnel have been underway.

The trip was an eye-opener to the 202 coal men who made it, acquainted as many were, with underground operations.

Among those in the party were P.M. Snyder, president of the C.C.B. Coal company, Mt. Hope; T.H. Snyder of the Sterling Coal company; P.C. Graney, general mgr. Of C.C.B; Wm. E. Tissue, sales manager of White Oak Coal company; E.R. Lynch of Gulf Smokeless Co., Wyco; Charles McMorrow, safety director at Eccles; Dr. B.B. Richmond of Skelton; Dr. D.L. Hill, Mabscott; Robert Lilly, president of New River & Winding Gulf Mining institute; Hobart Watson, Mt. Hope, secretary of the institute; W.R. Ballard of the McKell C&C Co.; W.M. Ward, superintendent at Kilsyth; George Dixon of the Price Hill Colliery Co.; J.C. Mabe, of the engineering department of C.H. Meade Coal Co.; Rev. E. Gibson Davis, pastor of the Baptist church, Beckley; Charles and Edward Howard of the New River company, Mt. Hope; J.S. Mason, superintendent at Willis Branch; E.H. Graff, safety director for the New River company, Mt. Hope; William Garvey, general manager of the Maryland New River company.

COLORED GAULEY BRIDGE MAN VICTIM OF SILICOSIS
Fayette Tribune, February 22, 1933

Samuel Ward, 30 years old, a former tunnel worker, died last Wednesday at his home at Gauley Bridge after an illness reported to be silicosis, contracted in working in the New river tunnel project at Hawks Nest several months ago.

An autopsy was performed here and physicians fixed his death as resulting from silicosis, a disease peculiar to workers in silica sands and rock.

Funeral services were held at Vanetta at 2 o'clock. Seven brothers and sisters survive. He was unmarried.

18 NEW SUITS ARE FILED AGAINST CONSTRUCTION CO.
Fayette Tribune, March 1, 1933

Eighteen additional suits were filed late Saturday against the Rinehart & Dennis and New Kanawha Power company, constructors of the dam and power projects and tunnel at Hawks Nest. All of the suits were for trespass; damage of $25,000 each.

Those who filed suits include Arthur Cox, Wilsie Cole, Charles Butler, Wayne Eastman, Robert Nichols, Edward Hall, A.L. Britt, Ballard Blankenship, James Colmore, Jessie Charles, Roy Kirk, Vernon Eastman, William Boggs, William Mosby, James Simms, Charles Campbell, Elijah Johnson and E.C. Murphy.

With the filing of these 18 suits the total number of suits against the defendant companies is now brought up to 111 suits. And the total damages sought is $2,725,000. Something like 12 suits are entered for $10,000 damages, while the other 99 are for $25,000 and $50,000. Those suing for death due to inhaling the silica dust are for $10,000 while those who have contracted silicosis and still living are suing for personal injury.

FIVE MORE SUITS ENTERED
Fayette Tribune, March 8, 1933

Five suits were entered Wednesday against Rinehart & Dennis Co. and New Kanawha Power Co by Ed. More, Frank Hugson, Albert Tate, Esten Janey and Willie Purcel. These were for $25,000 damage each.

This now brings the total number of suits filed against these defendants up to a total of 130 asking damages totaling over $3,000,000.

FIRST SILICOSIS CASE TO BE HEARD IN COURT THURSDAY
Fayette Tribune, March 15, 1933

[The first paragraph of this article dealt with unrelated issues and was therefore omitted.]

Thursday has been set as the time for hearing of the case of Dora Jones, admrx. of the estate of her late husband, Cecil L. Jones against the Rinehart & Dennis and the New Kanawha Power company. Jones is said to have contracted silicosis while working in the tunnel at Hawks Nest, that is being constructed for the New Kanawha Power company, through which the water of New River will flow to generate power at Gauley, and which is to supply power for plants in the New River valley.

The suit is for $10,000 and was recently argued as a test case before the supreme court which ruled that employes or administrators only sue the defendant companies. Over 200 witnesses have been summoned in this case, 126 for the plaintiff and 75 for the defendants. Hubbard & Bacon, Fayetteville law firm will represent plaintiff while the defendant companies will be represented by Judge W.L. Lee, C.W. Dillon and Atty. E.W. Knight, of Charleston.

To date 12 post mortem examinations have been made by Dr. C.R. Harless of Gauley Bridge and C. Dodd of Fayetteville. These examinations it is said have proved that death was due to silicosis, a disease which causes a fibrous tissue to form on the lungs which in time, closes the air passages, according to the physicians.

It has been estimated that 130 have died from this disease and that 350 have contracted the disease and are still living.

50 ADDITIONAL JURORS DRAWN FOR CIRCUIT COURT SERVICE
Fayette Tribune, March 15, 1933

An additional list of 50 jurors were drawn in open court Monday to report for duty Thursday at which time the case of Dora Jones vs. Rinehart & Dennis and others will be take up. They are as follows:

T.G. Griffin, Lawton; E.L. Gwinn, Springdale; W.C. Tincher, Crickmer; T.A. Dotson, Meadow Bridge; B.C. Goins, Backus; Lannie Armstrong, Crickmer; W.V. Hess, Gauley Bridge; Joe. Skaggs, Romont; N.P. Humphrey, Gauley Bridge; Charles Gilmer, Glen Ferris; B.H. Youell, Boomer;

D. H. Dunbar, Jodie; C.A. Hodger, Bellwood; J.L. Martin, Danese; L.D. Parker, Maplewood; S.F. Sanford, Bellwood; R.S. Aldridge, Corliss; R.L. Saunders, Sewell; Earl Holliday, Winona; J.R. Dietz, Hico; R.O. Nutter, Lookout; Robert Dice, Nuttallburg; W.F. Baber, Hico;

Henry Howell, Divide; A.F. Bell, Victor; E.E. Straughan, Ansted; James Sowder, Ansted; M.B. Ellison, Lansing; Mack Gill, Ansted; Clarence Mohr, Victor; J.J. Turner, Montgomery; G.S. Kelly, of Page; J.E. Kincaid, Jr., Kincaid; M.S. Bowles, Montgomery; B.J. Wriston, Kingston; Warren Beale, Powellton; H.W. Johnson, of Kincaid; Emery Kincaid, Kincaid; J.W. Mahoney, Scarbro; Charles Ongaro, Kanawha Falls; John White, Montgomery; George Myles, Fayetteville; J.D. Shultz, Fayetteville; Robert Massey, Red Star; R.E. Amick, Fayetteville; J.T. Grose, Fayetteville; Tim Blake, Oak Hill; Ben Hurvitz, Mt. Hope; W.R. Feazell, Mt. Hope; W.H. Reid, Scarbro.

FIRST SILICOSIS CASE GETS UNDER WAY IN COURT
Fayette Tribune, March 22, 1933

The first of 145 suits filed against Rinehart & Dennis Company, Charlottesville contractors, for the three and one half mile tunnel from Hawks Nest to Gauley, got under way in circuit court Thursday afternoon. Twenty-nine jurors were drawn before a jury of 12 could be empanelled. The jury as finally selected after three hours of questioning consisted of Mack Gill, Ansted; T.A. Dotson, Meadow Bridge; Emery V. Kincaid, of Kincaid; S.F. Sanford, Bellwood; J.E. Boley, Victor; W.H. Reid, of Scarbro; James Sowder, Ansted; B.C. Goins, of Backus; G.H. Berry of Backus; Robert Dice, Nuttallburg; A.F. Bell, Victor, and R.T. Saunders, of Sewell.

Occupations of the jury selected include 2 farmers, 4 miners, 1 merchant, 1 electrician, 2 mechanics and 2 connected with the C&O Railway Company.

Raymond Johnson, 37, of Gamoca, is the plaintiff in this first suit. He is suing for $25,000 damage. He worked from March 1931 to March 1932 in the tunnel as a steel jacker, and a driller and is suing for disability due to silicosis, alleged to have been contracted while engaged in this work.

Jones Case Halted

The case, which was first set to be tried during the special term of the Fayette county circuit court, was that of Dora Jones, admrx. of the estate of her husband, Cecil Jones, was halted because of a rule of prohibition by the supreme court. The defendant company had filed a petition for a rehearing of the case.

Legal talent connected with this case for the plaintiff includes Attorneys F.N. Bacon, Frank Love, W.E. Teubert and A.A. Lilly. For the defendant, Atty's C.W. Dillon, W.L. Lee and George Couch of Charleston connected with the firm of Brown, Jackson & Knight.

Opening statements for plaintiff were made by Atty. F.N. Bacon and for the defendants, Atty. C.W. Dillon. In his opening statements Atty. Bacon cited seven counts where plaintiff would prove the defendant company negligent in regard to ventilation in the tunnel drilling and failed to provide safe working conditions; also that Chief of the mining department, R.M. Lambie was first refused admission to inspect this tunnel and upon an order by the attorney general stating that the tunnel came under the jurisdiction of the mining department because of the commercialization of the silica sand, and the tunnel would fall under the classification of a sand mine, and that Chief Lambie should inspect same; and also that the defendant company had a system of signals and watchmen out to warn those in charge when mine inspectors were about to inspect the works; that the company in order to get the work along faster, did not use water in the drilling and thus the drillers inhaled quite a bit of silica dust, which permeated the tunnel at all times.

Dillon Denies Charges

Atty. C.W. Dillon stated that the contractors for this tunnel had been in the contracting business for the past 30 years and that this tunnel was the best ventilated ever to be under construction by the Rinehart & Dennis company. He also stated that by witnesses they would prove that the defendant company was not negligent in providing proper ventilation in constructing this tunnel.

Plaintiff has summoned 175 witnesses in the case while the defendant company will have 75 witnesses.

[section omitted]

It is alleged the drilling went through silica stone having a percentage of 99 of pure silica which is almost pure glass. The plaintiff in this first suit is white and was a former resident of Nicholas County. He worked in what is known as Project No. 1 near Gauley Bridge.

[remainder of article illegible]

COMPLETE SECOND WEEK OF TESTIMONY IN SILICOSIS CASE
Fayette Tribune, March 29, 1933

Wednesday completed the second week of testimony in circuit court in the case of Raymond Johnson, of Gamoca, who is suing the Rinehart & Dennis company and E.J. Perkins for $25,000 for damage he alleged is due to having contracted silicosis while working in the Hawks Nest underground tunnel project.

Early Wednesday afternoon Dr. L.R. Harless, of Gauley Bridge, was placed on the witness stand to give medical testimony. Dr. Harless told the jury that he had made a special study of silicosis during the past six or eight years. The disease, he said, had been known for centuries, but only during the past 25 years as silicosis. He testified that he had examined 150 to 175 men who had or were alleged to have silicosis during the past two years. Ninety-five percent had silicosis, he said. On examination, he stated that Raymond Johnson had been under his observation and treatment for the past year and that in his opinion, the disease was incurable and that Johnson could not live more than a year.

Nine Autopsies Performed

Dr. Harless told the jury that he had performed nine autopsies of the lungs of men who had worked in the tunnel and are now dead. He exhibited lungs of Cecil Jones and Louis Walter Street, alleged to have died of silicosis.

Several other expert medical witnesses will follow Dr. Harless on the stand before the plaintiff's case is completed. They include Dr. W.R. Huey and Dr. Lambert, of Charleston, and Dr. Hayhurst, of Columbus, Ohio.

Wednesday morning, Frank Gee, of Fayetteville, and the plaintiff himself were on the witness stand. Gee was a foreman for Rinehart & Dennis for some time and told of the working conditions in the tunnel in detail. Johnson's testimony was a summary of conditions described for the past several days by a half hundred workmen. Stripped to the waist early Wednesday afternoon Mr. Johnson exhibited his bare chest to the jurors as to the external features. He stated that he was 37 years old and now weighs 143 pounds as compared to a normal 175 pounds. He described the symptoms of the disease and the mental agony which he attributed to his condition.

To Finish This Week

Counsel for plaintiff intimated that the presentation of their side of the case might be completed this week. Due to the peculiar case, progress has been rather slow. Testimony during the week has related to working conditions which the plaintiff alleges were unhealthy and detrimental to lives of the men. Witnesses sought to show that Rinehart & Dennis were negligent in providing suitable working conditions.

Jones Back on Stand

During the course of the trial, Owen M. Jones, chief engineer for the New Kanawha Power company, on Tuesday was placed back on the witness stand for further testimony. Ten days ago he was interrupted by news of the death of his father. His testimony related to the relation of the companies and disposal of the silica rock. Some 300,000 tons of the stone were shipped to Alloy for future use for the manufacture of ferro-silicon.

Admits Lambie Letters

Over the objection of defense counsel, correspondence of Chief Lambie of the department of mines and Alderson Simms, an inspector, were read into the records by F.N. Bacon, chief counsel for plaintiff. The letters related to working conditions in the tunnel and instructions for correction of the condition.

Dr. R.C. Mitchell of Eccles, former company physician, testified Tuesday as to the number of men who became sick while working in the tunnel. One hundred or more had been examined by him and sent to the hospital he stated. The exact nature of the sickness was not known when it was first discovered, the doctor said.

Neighbors Testify

A dozen or more neighbors of Johnson when he resided in Nicholas county and later at Gamoca were on the witness stand during the course of the week. They told the jury that Johnson was a fine specimen of manhood and in excellent health prior to working in the tunnel.

DEFENSE OF SILICOSIS CASES BEGUN IN COURT HERE
Fayette Tribune, April 5, 1933

Rinehart and Dennis company started Monday to build up a defense in the first of some 150 cases involving alleged damages to workmen who contracted silicosis while working in the New Kanawha Power company tunnel on New River at Hawks Nest. Attorneys for Raymond Johnson, plaintiff in the first case, rested their case Monday afternoon after more than two weeks spent in an attempt to prove the defendant's negligence in providing safe working conditions for their workmen.

Owen M. Jones, chief engineer for the New Kanawha Power company, was the first witness for the plaintiff. Mr. Jones spent several hours on the stand. He explained in detail the plan and method of construction of the tunnel to provide a foundation for later testimony. He testified that working conditions were safe and that he himself spent a great deal of time in the tunnel in connection with his duties.

Mr. Jones on examination stated that he would rather breathe the air in the tunnel than in the court room, and that it caused him less discomfort. Visibility was good, and as far as he knew the tunnel did not contain impure air, gases or fumes and little dust. The work was done scientifically and according to approved practices, he said.

On cross examination, Mr. Jones admitted getting letters from Chief R.M. Lambie, of the department of mines relative to unfavorable inspections, but stated that he merely smiled at them. The tunnel, he said, was build under authority of the West Virginia public service commission under the water power act.

Two tests were made as to the purity of the air by a reliable chemist, he testified.

Foremen Testify

C.C. Waugh, superintendent of the No. 1 project and Charles M. Gilmore, a heading foreman, followed Mr. Jones on the stand. Mr. Waugh told the jury that he missed only five days of work and worked 12 to 14 hours a day on both shifts, testified that he had worked 35 years in tunnels and the working conditions and machinery in this tunnel was the best he had known. He said the drills were operated with water at all times and silica rock was wet at all times and there was plenty of fresh air in the tunnel and that there was little or no dust.

He contradicted statements of plaintiff's witnesses that he had signal systems to warn men when mine inspectors came and denied firing men for drilling with water.

Gilmore corroborated Waugh's statements in every respect. He denied that he slept on the job every night or was ever intoxicated on the job as witnesses for the plaintiff testified.

Wednesday morning, some half dozen former employees of the company told the jury of the excellent working condition in the mine and the fresh air. Air came through the 24 inch vent tube with such force that hats were blown off the men's heads when near the end in the heading, the men testified.

Among witnesses was Henry Abernathy, chief electrician for Rinehart & Dennis, his assistant, Sam Branham, John Kistler, a shovel operator, Charles Russell and Will Robinson, the latter two colored laborers.

It is not known how much more time will be spent in the trial, but it is expected that another week or ten days will be required.

X-RAY EXPERT SAYS JOHNSON HAD TUBERCULOSIS
Fayette Tribune, April 12, 1933

Testifying as an expert medical witness for the defendant, Dr. Henry K. Pancoast, professor of the University of Pennsylvania at Philadelphia, today told the jury in the damage suit of Raymond Johnson against Rinehart and Dennis that in his opinion, Mr. Johnson did not have silicosis. Medical witnesses for Johnson two weeks ago testified that Johnson had silicosis in the latter part of the second or first part of the third stage after physical examination and x-ray examinations.

Dr. Pancoast made his statement after a thorough study and examination of two x-ray films made of Johnson's lungs. He interpreted the shadows on the film to the jury and stated that Johnson's disease was shown by the film was an advanced stage of tuberculosis.

Silicosis is a symmetrical disease in contrast to tuberculosis when interpreted by x-ray films, he said. Dr. Pancoast is a recognized specialist in radiology which he has studied since 1902. He testified that he had examined some 140,000 x-ray films. Of these two to three thousand were silicosis suspects.

Lambie Says Dust Negligible

On the stand all day Monday, Robert M. Lambie, former chief of the state mining department, testified that the Hawks Nest tunnel was practically dust free. He listed inspections made personally in 1931 and told the jury that men were easily distinguishable from 500 to 700 feet away; that drills were operated with water and that the air in the tunnel's ventilating tube had a velocity of 27 miles an hour. Lambie appeared as a witness for Rinehart & Dennis company.

Several letters that Lambie wrote concerning the tunnel were read into the record. They were at variance with his testimony of direct examination. The former mining chief, under cross examination as to the different letters, said he has been misinformed by his inspectors concerning conditions. He mentioned that he had recommended the use of respirators and other safety features, but later withdrew them verbally after conferences with the contractors.

Members of representatives of three large contracting firms were presented Tuesday as witnesses for the defendant. Prior to this, L.T. Nuckols, a C&O assistant division engineer of Ashland, Ky., told the jury of his visit to the tunnel and Mr. Lambie completed his testimony.

H.W. Nelson, of Princeton, N.J., a member of the firm of Nelson and Chase Gilbert company, told the jury that in 26 years experience in which 12 tunnels were among jobs completed, he had never heard of the disease of silicosis, a disease peculiar to workers in silica dust. His company bored three tunnels on the Virginian railroad at Gilbert.

C.M. Faulkner, New York city assistant superintendent of A. Guthrie and Company, which built the Nicholas, Fayette and Greenbrier railroad line between Gauley Bridge and Nallen, and L.T. Chisolm, a contractor of Louisa county, Virginia, also testified that they had never heard of the disease. Each of the three men stated that they had visited No. 1 tunnel project while it was being constructed. The equipment was of the best standard, drills were operated with water, ventilation and air conditions were excellent and no dust was noticed, each testified.

J.W. Eggleson of Philadelphia, a representative of the Goodall Rubber company, testified as to the amount of air and water hose his company has furnished for use in Tunnel No. 1. He also told of good working conditions in the tunnel while he was making inspection of hose.

Tuesday afternoon, C.B. Bishoff and D.R. Sullivan, mine inspectors of the state department of mines were witnesses. Mr. Bishoff said he had made several inspections of the tunnel as official business. The drills were operated wet, the air was good and there was little or no dust visible he testified. Asked about certain recommendations he had made to improve conditions, he stated they were mere precautionary instructions and were never violated as far as he knew. Sullivan's testimony was similar to that of Bishoff. He said that officials of Rinehart & Dennis company had never failed to cooperate with him at any time.

DISAGREEMENT IN SILICOSIS TRIAL; JURY IS DISMISSED
Fayette Tribune, April 26, 1933[75]

After 20 hours deliberation, at the end of which time they were reported standing seven to five for the plaintiff, but hopelessly deadlocked at that point, the jurors in the first silicosis trial in circuit court were discharged by Judge Eary Monday evening. The suit for $25,000, instituted by Raymond Johnson against Rinehart & Dennis, Inc. and New Kanawha Power Company, had been in court more than a month.

No announcement was made as to whether the case would be retried, or whether any of a number of similar cases would be taken up later. Approximately 150 suits have been filed.

Testimony on vital points in the case conflicted sharply.

The case goes on record as being the longest ever tried in Fayette county. It began on Thursday, March 16 and the testimony was completed on Thursday, April 20. From the time the case began until the jury was dismissed it was 41 days. A total of 169 witnesses was heard.

---

75. Appendices. "Defendant's Instructions to the Jurors," pg. 207.

[several paragraphs omitted]

Counsel for the plaintiff in the silicosis case were F.N. Bacon, of Fayetteville; A.A. Lilly, Charleston, and W.E. Teubert, Ashland, Kentucky. For the defendants, C.W. Dillon and [J]udge W.L. Lee, of Fayetteville, and George Couch, of Charleston.

The jury was composed of the following: R.T. Saunders, Sewell; A.F. Bell, Victor; Robert Dice, of Nuttallburg; G.H. Berry, Backus; B.C. Goins, Backus; James Sowder, Ansted; W.H. Reid, Scarbro; S.F. Sanford, Bellwood; Emery Kincaid, Kincaid; T.A. Dodson, of Meadow Bridge, and Mack Gill, of Ansted.

TENTATIVE DATE IS SET FOR "SILICOSIS TRIAL"
Fayette Tribune, May 17, 1933

The widow of Cecil Jones will be the plaintiff in the next "silicosis trial," according to a definite announcement made at noon today by F.N. Bacon, of counsel for the plaintiff. Jones was a resident of Gamoca. Depositions were being taken today of witnesses in the case. It was also stated that testimony used in the first hearing will again be employed.

June 5 has been tentatively designated as the date on which the next "silicosis case" will come before Judge J.W. Eary of the circuit court, it was learned on Tuesday. Whether the session will be called as a special one or a continuance of the April term was not definitely learned.

A bill designed to relieve the local situation where approximately 150 suits have been lodged against Rinehart & Dennis, Inc., was killed in the house of delegates at Charleston, Monday. The measure proposed that courts be permitted to consolidate suits and try them as one upon petition of two or more plaintiffs. Had the measure been approved, it would have had the effect of expediting the hearing of the main claims.

[last section omitted]

SECOND SILICOSIS CASE POSTPONED TILL JUNE 12
Fayette Tribune, June 7, 1933

The second silicosis trial which was to have got under way on Monday morning, has been postponed one week in order to secure some judge to either hear the case here, or take Judge Eary's place at the June term of the Nicholas circuit court at Summersville.

It is expected that some one will be secured in the near future. It was thought that Judge Miller of Spencer would be able to hold one of these courts but because of ill health, he will not be able to serve.

The second case is that of Dora Jones, administrator, who is suing the Rinehart & Dennis company for $10,000.

The plaintiff in her suit alleges that her husband died from having contracted silicosis while engaged in working in the Hawks Nest tunnel through which New River will be diverted to supply power for various plants in the Kanawha valley.

The jurors who will report for duty next Monday are J.W. Arthur, Gauley Bridge; B.M. Brown of Carbondale; Lewis Bostic, Meadow Bridge; W.M. Beckett, Page; Fred Bush, Ansted; Arnold Burgess of Oak Hill; C.W. Campbell of Smithers; J.C. Campbell, Russelville; W.L. Diehl, Meadow Bridge; Thos. Epps, Winona; J.E. Gwinn, Gauley Bridge; Howard Hess, Lucas; U.S. Jones, Fayette; R.L. Long, Powellton; Chester Legg of Fayetteville; J.L. Meadows, Edmond; Isaac Murray, Oak Hill; G.F. Marshall, Kilsyth; R.L. Maynor, Fayetteville; W.L. Neeley, Crickmer; C.N. Proctor, Clifftop; W.A. Robertson, Sewell; G.T. Smailes, Landisburg; C.E. Vawter, Ansted; W.L. Wriston, of Kincaid.

MORE SUITS FILED AGAINST HAWKS NEST CONTRACTORS
Fayette Tribune, June 7, 1933

Six more suits were filed Friday on pauper's affidavits against the New Kanawha Power Co. and Rinehart & Dennis company and others for $25,000 damage each. These suits were instituted by Gilbert Crumpton, James Gollman, Jasper Tobee, Robert Johnson, J.B. Murray and Curlin Pitts.

These plaintiffs will be represented by Atty. C.R. Summerfield.

This makes a total of 156 suits pending in circuit court against the defendant companies, totaling in the aggregate about $5,000,000 in damages.

POST MORTEM PERFORMED ON BODY OF HAWKS NEST MAN
Fayette Tribune, June 14, 1933

A post mortem examination was made on the body of Fred Chattfield, colored, last week at Dodd's undertaking establishment here, the heart and lungs being removed to be used as evidence in some of the silicosis suits which have been filed in the local circuit court.

The examination was made by Dr. L.B. Harless of Gauley Bridge.

Chattfield had entered a suit against Rinehart & Dennis, Inc., seeking $25,000 alleging that he had contracted silicosis while working in the Hawks Nest tunnel.

The body was shipped to Atlanta, Ga., yesterday for burial.

JURY IS DISCHARGED IN SECOND SILICOSIS CASE
Fayette Tribune, June 14, 1933

The jury panel drawn to hear the second silicosis case in circuit court, was discharged Tuesday evening by Judge J.W. Eary without having been called upon to serve in the actual hearing of the case.

The jurors were summoned to appear Tuesday morning and when court convened it was found that only 11 of them were present. Court adjourned to meet again at 1:30. Meanwhile, it developed that attorneys for plaintiff and defense were in conference, presumably seeking a settlement of the controversy, so the jurors were dismissed until called.

Calling them again shortly after five o'clock, Judge Eary dismissed them without commenting on the status of the case.

While many rumors have been circulated on the present status of the case, no official announcement has been made by counsel on either side. The rumors being heard are believed to be without foundation, and no authentic source has revealed what transpired during the conference or any of the results thereof.

The case was not officially continued until the July term, the dismissal of the jury by Judge Eary not indicating that any such agreement had been reached

SILICOSIS CASES SETTLED: $130,000 IS PAID TO PLAINTIFFS
Fayette Tribune, August 2, 1933

Settlement of the so-called "silicosis" cases, growing out of the construction of the big Hawks Nest water tunnel, for the lump sum of $130,000 has been effected, 146 men or their estates have been compensated and individual payments have been made, or will be made within the next day or two, it was learned authoritatively this week.

While is has been generally conjectured for some time that a settlement had been made in these cases, no authoritative statement had been procured prior to this time.

The completion of these cases solves apparently what has been among the most widely discussed problems of the medical profession, as well as on construction workers of this type of project. Nationally recognized physicians, and many of them specialists in lung diseases, were brought into the case and the data obtained is said to be of extreme value to them.

A number of them including Dr. Hayhurst of Columbus, Ohio, who appeared as a witness for the plaintiff during the case of Raymond Johnson against the contractors, were in this section last week compiling the nearly completed data for the Medical Journal publication. The dissemination of this information will not only be for the benefit of the medical profession, but will also provide a guide for contractors in future projects which require removal of silica rock.

Those suffering from th[i]s disease as a result of their work at Hawks Nest will be compensated from the lump sum paid in ratio to their present disability. Of a total of 347 examined, 146 will receive compensation. The others were found not to be suffering from the disease. The examinations were conducted by Dr. Harless at Gauley Bridge, and minute records kept of each case.

Payments to individuals or their estates will range from $1000 downward. The payments are being made in lump sums.

Three general classes have been outlined by the examining physician. They are:
1. All positive first state silicosis cases: 31 men
2. All first stage acute silicosis cases: 79 men
3. All second and third stage silicosis cases and deaths: 36 men

A total of 157 suits had been entered against Rinehart and Dennis, contractors on the tunnel work and the sum of more than $3,000,000 was asked. Only one case ever came to trial, that of Raymond Johnson and after some six weeks of presentation of testimony and evidence, the case resulted in a disagreement of the jury. A second case, scheduled some two months ago, was not brought to trial, the jury being dismissed by Judge J. W. Eary following conferences with counsel for both sides.

Shortly after these conferences, and the subsequent dismissal of the jury, a trust deed in the amount of $130,000, of Rinehart and Dennis, was entered in the court house records, and then it was generally presumed that the sum was the amount called for in the settlement.

Attorneys for the plaintiffs who were present at the final settlement were W.E. Teubert, Gauley Bridge; F.N. Bacon, Fayetteville; J.K. Edmundson, Fayetteville; C.T. Dyer, Montgomery; A.L. Russell, Fayetteville, and A.M. Mason, of Charleston.

The defendant company was represented by the law firm of Brown, Jackson & Knight, Charleston; Judge W.L. Lee, Fayetteville, and Dillon, Mahan & White of Fayetteville.

DISMISSAL OF SILICOSIS CASES MEETS OBJECTIONS
Fayette Tribune, August 23, 1933

A motion was made to Judge J.W. Eary of the circuit court the past week to set aside the dismissal of 88 silicosis suits which were reported settled when the defendant company made a final settlement a month ago. Plaintiffs in these motions are represented by Attorneys Thos. C. Townsend, Ben Moore and E.S. Bock of Charleston.

These motions were made upon the following grounds: "That the said orders of dismissal, respectively, were entered pursuant to a purported settlement which purported to include the said several plaintiffs named along with a large number of other plaintiffs in similar cases against the same defendants; whereas in fact none of the respective movants received anything whatsoever by way of settlement of said respective cases, and has in fact made no settlement with the defendants or any of them, nor did any of the said plaintiffs order the dismissal of their respective actions for damages against Rinehart and Dennis and others."

[list of plaintiffs omitted]

The order was docketed and is under consideration at the present time by Judge Eary to be disposed of at a later date.

1934

MORE SILICOSIS SUITS ARE FILED
Thirty Additional Suits for $25,000 Each are Filed in Fayette
Montgomery News, Montgomery, W. Va., Thursday, January 4, 1934

Thirty additional silicosis suits were filed Tuesday afternoon in the office of the circuit clerk by the law firm of Townsend, Bock and Moore, of Charleston. The plaintiffs in these suits are seeking to recover $25,000 damages from the contractors of the three and one-half mile tunnel which was constructed between Hawks Nest and Gauley Bridge by the Rinehart and Dennis Company, Inc.

The plaintiffs included: John Austen, Edgar Bostic, Richard Boyd, Johnnie Bryson, Major Chattfield, Dewey Clemons, A.A. Cole, Earl Cole, Howard Cole, John Cole, George Cole, Ben Geither, J.T. Geter, John Gill, Jessie Gladden, Dock Gladden, Frank Gladwell, Henry Gibson, Steve Hayes, Johnnie Holmes, Henry Jackson, Jim Jordan, Cornelius Jumper, Clarence Meadows, Joseph Moore, Walter Raleigh, Fred Stinson, James Tillman, Horace Tidwell and James Watson.

SILICOSIS CASE TO COME UP MONDAY
Second Suit is to be Tried Before Judge Eary; $50,000 Damages Sought
Montgomery News, June 14, 1934

The second silcosis suit to be tried in the Fayette County circuit court is scheduled to get under way Monday morning, June 18, when the suit of the plaintiff Donald J. Shay vs. the Rinehart Dennis Company, a corporation, The New Kanawha Power Company, O. M. Jones, and E.J. Perkins will be tried before Judge J.W. Eary.

The plaintiff in this suit is suing the defendants for $50,000.00 This is one of the suits which was not dismissed in the general settlement of the many silicosis suits which was made last summer, when the defendants compromised with some 130 plaintiffs.

The suit of Raymond Johnson who sued the defendants for $25,000.00 was the first silicosis case to be tried in the Fayette County circuit court and resulted in a hung jury. The trial of the first suit lasted over five weeks in which some 200 witnesses testified in the case. Johnson was represented by the Attorneys F.N. Bacon, A.A. Lilly, of Charleston, and W.E. Teubert of Gauley Bridge. The defendants in the first suit were represented by Attorney George Couch of the law firm of Brown, Jackson and Knight of Charleston; Attorney W.L. Lee and the law firm of Dillon, Mahan and White.

The plaintiff in the second suit will be represented by the law firm of Townsend, Bock and Moore of Charleston while the defendants will be represented by the same attorneys who defended them in their first suit.

FIFTY-ONE SUITS FILED THIS WEEK
Silicosis Suits are Brought as Hearing of Case is Started
Montgomery News, June 31, 1934

Twenty-six additional silicosis suits were filed in the office of the circuit clerk Wednesday morning, June 20. The suits were instituted by Attorneys S.H. Ballard and W.W. Wertz of Charleston. The plaintiffs in each of these suits is suing the defendants for $25,000 damages. This brings the total up to 51 silicosis suits within the past week seeking damages of over one million dollars. The plaintiffs in the suits include: Sam Sadler, Jim Harris, Will Brown, Percy Shepherd, Robert Wilson, James Wilson, Melvin White, Wilbur Armour, Ferry Williams, George Watkins, James Harris, Fred Green, Robert Lynch, Willie Palmer, Fred Henderson, Ernest Hopson, Willie Davis, Ivory Marby, Williard Lee, Dave Coleman, T. Oliver, Ed. Foster, George Bradford, Monroe Johnson, Mose Jenkins and Eugene Warren.

The defendants names in the suits include the Rinehart and Dennis Company, Inc., The New Kanawha Power Company, a corporation, E.J. Perkins, O.M. Jones, and P.H. Faulconer.

The other 25 silicosis suits were filed in the office of the circuit clerk this past week on paupers affidavits by attorneys W.H. Haynes and Frank H. Brazie who will represent the plaintiffs in the suits against the builders of the tunnel.

Each of the plaintiffs in the suits are suing to recover $20,000 damages. They include: Charles Hartman, Lonnie Chapman, John Bell, Josh Crawford, James Bryson, Calvin Pugh, Fred McCoy, Washington Simmons, James Russell, Charles Kidd, Will Robinson, Williams Armour, John Johnson, Harvey Moses, Sam Yates, George Head, Horace Coleman, Ernest Brown, Ernest Hopkins, George Brown and Garten Collins.

SILICOSIS JURY FAILS TO REACH AGREEMENT
Montgomery News, July 12, 1934

Failing to reach an agreement, the jury which heard the case of Donald J. Shay vs: New Kanawha Power Company and Rinehart and Dennis Company was discharged by Judge Eary shortly after 10 o'clock Wednesday morning. The case was given to the jury Monday evening shortly before 5 o'clock. It was reported that the jury stood 10 to two during almost the entire deliberation, although it was not announced which side had the majority.

Shay asked $50,000 damages, alleging he contracted silicosis while working in the Hawk's Nest water tunnel in the silica sandstone. His was the second of the cases to be tried, that of Raymond Johnson also previously resulted in a hung jury.

SILICOSIS SUITS UP FOR ACTION
Plaintiffs Ask Right to Try Silicosis Suits in Other Counties
Montgomery News, October 4, 1934

The question of whether approximately 200 suits asking damages totaling $5,000,000 shall be heard in the Fayette County circuit court or in other tribunals will be ruled upon by the supreme court.

[paragraph omitted]

The suits are directed against the New Kanawha Power Company which started the project; Rhinehart and Dennis Company of Charlottesville, Va., contractor; O.M. Jones, chief engineer for the New Kanawha; and E.J. Perkins, superintendent for the contractor.

Edward S. Bock of Charleston, counsel for the plaintiffs said circuit Judge J.W.

Eary, of Fayetteville, has certified a motion for removal of the suits to other courts to the supreme court for a ruling after the defense contended sufficient cause was not shown for removal.

Bock claims that in the view of the fact that only one suit could be tried each court term in Fayette County, it would take years to complete the cases.

KANAWHA GETS SILICOSIS SUITS
Supreme Court Refuses to Countermand Transfer from Fayette Court
Montgomery News, October 18, 1934

Transfer of some 60 silicosis cases from the Fayette to the Kanawha County circuit court has been ordered, in effect, by the state supreme court, which refused to review the action of Judge J.W. Eary of the Fayette Tribunal, in ordering the transfer.

[paragraph omitted]

Counsel for the plaintiffs contended that it would take years to try all of the cases in the Fayette County circuit court. Defense attorneys asserted sufficient cause was not shown for removal.

The actions involve claims for damages totaling approximately $5,000,000, attorneys for the plaintiffs said.

FIVE DIE, TWO ESCAPE AS POWER PENSTOCK BURSTS
Fayette Tribune, November 8, 1934

Five men, among them John W. Hall, of Montgomery, representing the constructing engineers at the power dam at Gauley Junction and Hawks Nest, met their death, and two men were rescued when one of the four penstocks leading from the tube connecting with the overflow tank or the main tunnel burst last Thursday morning at 10:15.

The other dead are: D.W. Haynes, 34, of 513 Main St., Charleston, a foreman; A.L. Kube, 51, of Rhodesville, VA; Ralph Stringer, 33, Cleveland; and Paul Sykes, 23, Gauley Bridge.

Two of the men, Hall and Haynes, were drowned while the others were killed outright. Stringer was caught beneath the sheet of steel, some six by ten feet, which burst from the side of the 14-foot penstock, and was badly crushed. Sykes and Kube sustained fractured skulls when they were apparently hurled against the sidewall of the pit.

Two men, J.H. Patterson, an engineer employed by the New Kanawha Power

Tunnel/penstock junctions interior, 1934. *Courtesy West Virginia State Archives.*

Company, and Cecil Gay, engineer for Babcock & Wilcox company, the concern in charge of the steel installation, were rescued. Both were treated at the Coal Valley Hospital at Montgomery.

The accident, according to an official statement, occurred during a test period. It was learned, too, that the testing had been about 90 percent completed.

The penstocks through which the water goes direct to the turbines after leaving the overflow tank are steel tubes, 14 feet in diameter. They are underground some 30 feet, the tunnels in which they are laid being cut in solid rock.

Each of the penstocks, according to unofficial information, is laid in a separate tunnel, although openings provide access for workers from one to another. Walkways are provided around the tubes being quite narrow in some places.

The penstocks are about 35 or 40 feet long, and branch out from the main tube until the ends connecting with the wall next to the turbines are some distance apart, it being estimated that it is about 40 feet from number one to number four.

The seven men in the pit were testing penstock No. 4, according to information believed to be authentic. Hall and Patterson were on top of this tube, while the other men were alongside it, some 30 feet from the bottom end.

The accident occurred without warning. The large slab of steel suddenly broke clearly from the tube, crushing Stringer beneath it and the other men were instantly engulfed in the water. The pressure was from 100 to 115 pounds, and approximately 2,000,000 gallons of water are reported to have been in the tubes.

The bodies of Sykes, Stringer and Kube were recovered shortly after the accident, according to our information, all of them being near the scene of the accident. Hall's body, the last to be found, was not removed until nearly 4 o'clock in the afternoon of Thursday.

The escape of Gay and Patterson borders on the miraculous. The former, standing beside the penstock observing the test, was washed through the trenches and carried back and forth for some distance, completely submerged in the water. He was pulled to safety by workmen when washed near the edge of the pit. The latter, on top of the tube, was thrown into the water also and while submerged happened to grasp a ladder leaning against the side of the penstock and pulled himself to safety. They were both taken to the hospital immediately after being rescued.

Details of the plan of construction have not been officially revealed and reports on them vary. It has been reported that there are five, rather than four penstocks and that they led directly from the mouth of the tunnel rather than the overflow tank. This newspaper cannot definitely state which of these reports is correct.

The penstock which burst had been subjected, as have all the others, to every kind of test before it was placed under the pressure. All the welded parts, electrically joined, had been x-rayed and found to be perfectly joined. The penstock, according to our information, did not burst on a welded part, all the joints holding securely.

It was also unofficially stated that the tube had previously been subjected to 125 pounds of pressure and was to have been tested to 140 pounds later.

A Mr. Maxwell, said to be chief contractor in the steel construction, is reported to have left the pit not longer than five minutes prior to the accident, and he assisted in the rescue work.

Other officials present also aided in the work, among them being O.M. Jones, chief engineer located at Glen Ferris and well known here.

Physicians called to the scene and who entered the pits in the hope that they might be able to help the trapped men, were Drs. Simmons and Keyser, plant physicians, and Dr. C.W. Stallard, of the Coal Valley hospital.

STATE TO TAKE ACTION IN HAWKS NEST POWER PROJECT
Fayette Tribune, November 22, 1934[76]

Attorney General Homer A. Holt[77] said Tuesday he would file an intervening petition for the state in the suit by the federal government for an injunction to halt the construction of the Hawks Nest power project.

The attorney general said the state "naturally is interested in its water power resources," but declined to say anything further on the state's position in the action.

The government asks a restraining order against the Electro-Metallurgical Co., the New Kanawha Power Co., and Union Carbide and Carbon Co., of New York. The suit filed by District Atty. Geo. I. Neal, charges that the dam will interfere with navigation in Kanawha river, where the government has carried out a series of new locks and dams.

The defendants challenge the constitutionality of the federal water power act insofar as it attempts to give authority to the federal water power commission to exercise control over New river which the companies say is not a navigable stream.

The companies admit a Federal license was not obtained for the power projects but assent a state license was obtained, and that is all that was required.

STATES ENTERED IN HAWKS NEST LICENSE ACTION
Moves to Keep Government From Acting Against State Power Development: Claim New River Not Navigable Waterway
Montgomery News, December 20, 1934

Challenging the constitutionality of the federal water power act insofar as it attempts to give the Federal Power Commission jurisdiction over non-navigable streams, the State of West Virginia Friday intervened in the government's suit to require federal licensing of a New River hydroelectric project.

Attorney General Homer A. Holt said in the state's petition that the federal government "has no right, power or authority under the constitution of the United States to supervise, regulate or control or in any manner to interfere with the construction, maintenance or operation" of the Hawks Nest project in Fayette County.

The government filed suit June 11 to require the Electro-Metallurgical Company, the New Kanawha Power Company and the Union Carbide and Carbon Company, of New York, to obtain a federal license for the Hawks Nest project. It asks an injunction

---

76. Appendices. "Memorandum of The Public Service Commission of West Virginia to The Federal Power Commission," pg. 212.
77. See "Biographical Information," pg. 191.

halting construction until a federal permit is obtained, asserting no government agency authorized the project.

The defendant companies deny the government's claim that the New River is navigable, but assert it is non-navigable, that the government has no control over it, and attack the constitutionality of the water power act in attempting to give the power commission jurisdiction over the system.

Preliminary issues will be argued January 7 in Charleston before Federal Judge George W. Mc*[illegible]*. The attorney general's petition sets forth the state owns water power rights in its streams, that the defendant companies have complied with state statutes and have a state permit for the project, and asserts the state has exclusive jurisdiction over the stream.

The petition says if the court should require a federal license, it "would establish a precedent of federal ownership and control of hydroelectric rights, not only in the New River, but in many other streams in the state of West Virginia which are non-navigable and destroy the rights of the state to license, dispose of or use hydro-electric or water power rights in such streams."

Power Not Authorized

"It is an attempted assertion of powers not delegated by the constitution of the United States to the United States or prohibited by it to the states, but reserved to the states and constitutes an attempted encroachment upon the sovereign right of the state of West Virginia to own, use, control, or grant property and rights not belonging to the United States but solely to the state contrary to the fifth and tenth amendments to the constitution of the United States and provision of the state constitution."[78]

1935

SILICOSIS CASES SET FOR APRIL 8TH AT CHARLESTON
Fayette Tribune, March 7, 1935

Some 60 "silicosis" damage suits against the Rinehart and Dennis and New Kanawha Power Co., builders of the Hawks Nest hydro-electric project and tunnel which were transferred from the Fayette circuit to the Kanawha circuit court, and a test case was set for April 8th by Judge Arthur P. Hudson of the Kanawha Court, Saturday.

The test case which is likely to require some three or four weeks in trials, is that of the $50,000 suit of Donald J. Shay.

The other cases will be left on the open docket until the Shay suit is concluded.

---

78. Appendices. pg. 212.

Two of these celebrated cases for damages have been tried in the Fayette court which resulted in a disagreement of juries.

The cases were taken to the circuit court of Kanawha county after a ruling of the state supreme court that they could be transferred on petition of attorneys for the plantiff.

STATE ASKS DISMISSAL OF FEDERAL HAWKS NEST SUIT
Fayette Tribune, March 21, 1935

The state of West Virginia, the Union Carbide and Carbon corporation of New York, and its subsidiaries, the Electro-Metallurgical company and New Kanawha Power Co., have urged before the U. S. supreme court the dismissal of the federal government's suit to force the builders of the Hawks Nest hydro-electric project in this county to take out federal license.

The government contends that no federal license was obtained for the work, and that it has jurisdiction of the stream. The state and defendant companies attack the constitutionality of the federal water power act, insofar as it attempts to give the federal government authority over the stream.

The New Kanawha Power Co. obtained a state permit for the construction and transferred right to the Electro-Metallurgical Co., which has almost completed the project.

The case was taken to the supreme court by the government, after it had abandoned a suit in the federal district court at Charleston.

FAYETTE SILICOSIS CASE POSTPONEMENT IS ORDERED
Fayette Tribune, April 11, 1935

Postponement of the silicosis case of Donald Shay v. Rinehart and Dennis Construction Co., and others, in the Kanawha county circuit court, was ordered last Monday, pending a ruling on whether the cases have been outlawed by the statute of limitations.

Counsel for the defendants in the suit maintain that the suits, approximately 60 in number, should have been instituted within a year after termination of employment. Plaintiff counsel contend they were started within a year after the disease was contracted.

Damages approximating $1,500,000 are sought by the workmen who claim they contracted silicosis while working in the Hawks Nest tunnel.

SUPREME COURT TO DECIDE ON SILICOSIS CASE SOON
Fayette Tribune, April 25, 1935

That a decision will be rendered in the near future by the West Virginia supreme court regarding the status of the so-called silicosis cases, is stated in news dispatches from Charleston. The court has agreed to determine whether or not the cases have been outlawed by the statute of limitations.

The company in a test action docketed by the supreme court last Friday asserts the suits were not entered within the statutory time limit and cannot be maintained.

The test case is a suit by Lewis Scott for $25,000 damages. In that action the company asserts Scott has no suit because his claims were not entered until more than a year after his employment ended with the Rinehart and Dennis company.

HAWKS NEST CASE TODAY
Fayette Tribune, May 2, 1935

The arguments on the government's suit to require federal license for operation of the Hawks Nest hydro-electric plant were to be heard before the U. S. supreme court today. Attorney General Homer A. Holt headed the state's counsel.

STATE WINS HAWKS NEST FIGHT: NEW ACTION SEEN[79]
Fayette Tribune, May 23, 1935

The state of West Virginia won the first battle against the federal government in the Hawks Nest license argument before the U. S. supreme court Monday, but a new action to require federal license for the project is planned in the federal court, counsel has announced.

The supreme court ruled that there was no federal question presented.

The federal government seeks to force the Union Carbide and Carbon company to obtain a federal license to operate the project.

---

79. Appendices. "United States v. State of West Virginia, 295 U.S. 463 (1935)." pg. 223.

200 SILICOSIS CASES ARE OUTLAWED BY LIMITATION
Fayette Tribune, May 30, 1935

Approximately 200 of the silicosis cases were stricken off the record this week when the state supreme court at Charleston declared that the statute of limitation was applicable in that of Scott vs. the defendant companies. About that number of cases filed will come under the same statute, according to attorneys in the cases.

There will still remain, however, a considerable number of cases, it has been indicated. Whether they will be brought to trial within the near future has not been indicated yet.

1936

TUNNEL DEATHS IN HAWK'S NEST PROJECT CITED
* *New York Congressman Claims 476 Died, 1,500 have Silicosis*
* *Demand Investigation of Work Conditions*
* *Would have Federal Inquiry into All Tunnel Jobs*

Montgomery News, January 17, 1936

A staggering death toll among workers on the Hawk's Nest tunnel in Fayette County was cited Friday in Congress at Washington by Representative Marcantonio (R-NY).

Marcantonio said 476 have died and 1,500 are dynig of silicosis contracted while working on the Fayette County tunnel. He termed tunnel deaths "America's greatest industrial catastrophe."

The inquiry which Marcantonio would have made by the labor department would include all tunnel operations, and not only the Hawk's Nest project. He presented a resolution in the house Monday.

The Hawk's Nest tunnel, built to divert water from New river for a hydroelectric project, was completed in 1934.

Marcantonio said: "Four hundred and seventy-six men have died and 1,500 are doomed, because silicosis, a lung disease is incurable."

He charged that the bodies of 169 workers who died of the disease after working in the huge tunnel, cut through silica rock in the mountains one mile east of Gauley Bridge, "were dumped into a cornfield and their only grave stone were cornstalks waving in the wind."

He said he would ask the house labor committee, of which he is a member, to immediately consider his resolution and that a stockholder in the Union Carbide and Carbon Company, several persons suffering silicosis, and investigators would testify.

Marcantonio added he was "looking into whether there has been violation of the federal power commission regulations" in the construction of the $20,000,000 project.

The federal government, about a year ago, filed suit to halt construction of the Hawk's Nest power dam, charging it would interfere with navigation. The builders and the state of West challenged the power commission's authority to exercise control over the stream. The action, abandoned at Charleston, W. Va., later was instituted in the supreme court and dismissed there.

Scores of suits filed against the companies building the tunnel for damages running into many millions of dollars were settled out of court, and the West Virginia supreme court several months ago threw out over 200 suits on the grounds that they had not been instituted within the time provided by law.

The workmen and administrators of their estates charged negligence, asserting the workmen contracted silicosis by breathing fine dust in cutting through the silica rock, and that proper safeguards were not provided. The plaintiffs further said, in the recent cases, that they would not have brought suits sooner because they did not know they had contracted silicosis until many months after their employment terminated.

BIG HAWKS NEST PLANT TO BEGIN OPERATION SOON
Engineers Say Giant Power Firm Will be Ready to Open in June

WILL DIVERT WATER FROM NEW RIVER
Plant will Send Current 10 Miles to E.M. Company's Plant at Alloy
Montgomery News, January 24, 1936

The famous Hawks Nest Power Plant, center of litigation and investigations ever since construction began in 1930, is being prepared to begin operation in June.

Engineers inspecting the gigantic hydro-electric plant this week-end said they will be ready by the end of spring to divert water from the New River into the turbines through the 3 1/4 mile tunnel which now is being investigated by a congressional committee.

The plant, costing $8,000,000 and capable of developing more than 100,000 horsepower, will send the current 10 miles to the Electro-Metallurgical Company's plant at Alloy, W. Va., where ferro-silica is made for the steel industry.

Construction by Rinehart and Dennis, of Charlottesville, Va., began June 19, 1930, and in less than a year the contractors were involved in 538 suits brought by workers claiming they suffered from silicosis, a lung disease caused by inhaling particles of silica sand.

Tunnel intake during construction, showing the rack panels installed. *Courtesy West Virginia State Archives.*

The men claimed they contracted the malady while drilling the huge tube—31 feet in diameter—through a great deposit of almost pure silica. Rinehart and Dennis declared the suits cost them $300,000, and the silicosis cases now are being investigated again, this time by Congress. Representative Marcantonio, Republican, New York, asked the probe, with a declaration 476 men died of the ailment and 1,500 are dying.

In 1934, before the litigation with workmen closed, the Electro Metallurgical Company was in court defending the right to build the plant. The federal power commission, claiming the construction should have been licensed by the federal government instead of the State. Asked an injunction to stop the work. The State intervened, and in May, 1935, the supreme court of the United States threw the suit out, ruling a decision would have to be given in a district court before the high tribunal considers the matter.

The case has never been re-filed. Under terms of the permit granted by the State, the project which borders the State-owned Hawks Nest Park, can be reclaimed by West Virginia within 50 years at the original cost of construction.

Power house during construction. Generators Nos. 1 to 3 are completely erected. *Courtesy West Virginia State Archives.*

The New-Kanawha Power Company, a sudsidiary of the Metallurgical, is to operate it.

The plant is equipped with four 30,000 kilovolt amperes generators having a total output under normal water supply conditions of more than 100,000 horsepower.

The water, diverted from New River by a gigantic dam constructed under the Hawk's Nest scenic point in New River, drops 163 feet in traversing the tunnel and produces a head force of between 60 and 65 pounds per square inch at the turbine.

A governor on each unit opens and closes the water gates leading to each turbine as the speed varies from the normal of 150 revolutions a minute.

The generated 6,300 volts of electricity are stepped up by transformers to high 'voltages' for transmission to Alloy to supplement the power from the present steam plant.

The two largest products of the plant are ferro-silica, used in the steel industry as a method of forming slag to carry off impurities; and ferro-chromium, used in making tool steel.

Dam during construction, looking upstream, showing six completed spans of the operating bridge. *Courtesy West Virginia State Archives.*

TUNNEL WORKING CONDITIONS ARE GIVEN TO PUBLIC
P.H. Faulconer, President of Rinehart and Dennis Co., Reviews Tunnel Building

PRECAUTIONS USED TO SAFEGUARD WORKMEN
Recent Publicity Characterized as Gross Misrepresentations and Falsehoods
Montgomery News, January 24, 1936

P.H. Faulconer, president of the Rinehart and Dennis Company, Inc., contractors, has authorized the following statement:

"The press has requested this company to make a statement with regard to working conditions which prevailed during the construction of the Hawks Nest power development at Gauley Bridge, W. Va.

Power house site during construction, showing the excavation for the surge tank and backfill over the tunnel portal. *Courtesy West Virginia State Archives.*

"In view of the gross misrepresentations and falsehoods which has recently been circulated concerning this project, we hereby acquaint the public with the real facts. The entire record of the operation is an open book. We invite any interested person to scrutinize every step of the work, convinced that any impartial survey will establish beyond a question of doubt that all the precautions known to engineering science were employed by us to safeguard the workers on this power development.

"Work on the project was begun about April 1, 1930. The development comprises a dam at Hawks Nest, W. Va., a power house at Gauley Junction, about a mile upstream from Gauley Bridge, and a tunnel three miles long between the dam and the power house. The excavation of the tunnel was completed by the end of 1931 and the entire job virtually finished by the end of 1934.

"From the beginning of the development until its completion 4,948 men were employed, of whom 1,688 were white and 3,280 colored. Of this number the working force assigned to tunneling totaled about 2,500 of whom approximately 500 were white and 2,000 colored. The largest number employed at any one time on the entire

Excavation completed for west abutment and downstream portion of tunnel intake. *Courtesy West Virginia State Archives.*

project was about 400 white men and 850 colored in all, 1,250. The largest number at work on the tunnel at any given date was around 600, divided into two shifts and distributed at the four headings.

"Tunneling was done by two shifts daily, except Sunday, when no work was done. A shift lasted ten hours, or less. Pay was for the full ten hours even if the shift was shorter, which usually was the case. There was an interval of at least two hours between the cessation of work by one shift and beginning by the next. During this time both the ventilating and compressed air lines were continuously delivering fresh air into the tunnel.

"Every known device to protect the lives of workers was included in the complement of machinery assembled to carry on the operation. The tunneling equipment was new and carefully selected. It included the latest Ingersoll-Rand air drills equipped with standard water heads for wet drilling. Wet drilling, by which the creation of dust is prevented, was insisted upon at all times and there were ample pumps and piping to carry water to the drills. Air compressors to supply air from outside the tunnel to the drills were installed. We bought new electric shovels to load the spoil and new electric locomotives to shift cars at the headings.

"A thoroughly up-to-date system of ventilation of even great capacity than that recommended by experts was installed. There were new motor driven fans to supply ventilating air and 24-inch tubing of the most improved type to convey the air to the headings.

"An example of the extreme care exercised to provide proper ventilating is furnished in the size of the tubing used. Ventilation experts had recommended 20-inch tubes as entirely adequate to provide the proper amount of air, but we obtained 24-inch tubes in order to supply still more air. These tubes delivered 7,000 cubic feet of air per minute to each heading. In addition, the exhaust from each drill delivered more than 150 cubic feet of air per minute. In all there were over 8,000 cubic feet of air per minute delivered to each heading. This averaged in excess of 150 cubic feet per minute per man in the tunnel, at least 50 per cent more than required by the State mining law.

"Conditions inside the tunnel were as favorable as engineering science could make them, a fact commented upon by many visiting engineers and contractors. The air was cooler in summer and warmer in winter than outside. The expansion of air from the drills produced a coolness at the headings.

"Samples of air taken from the tunnel by chemists were analyzed and found freer from contamination by noxious gases than in Fifth Avenue in New York Ctiy. There was never any dust haze in the tunnel while the men were at work, as has been charged. The atmosphere was so clear that signts of a thousand feet and over were taken by the regular surveying parties in the tunnel whose work required great precision. Blasting was done as each shift completed its work and after the men had left the tunnel heading. In the interval between shifts the smoke and dust were cleared away by ventilation.

"There were other safeguards. At regular intervals the blasting work was inspected by experts and instructions given in the use of explosives. We had an arrangement with the Coal Valley Hospital in Montgomery for hospitalization of sick or injured employees. Two physicians, one resident on the work, were in attendance daily upon the men. The sanitation and cleanliness of the camps were under constant supervision.

"The great exaggeration of the mortality rate is shown by the following figures. By the end of 1932, when the bulk of the heavy work was completed, the official records at Charleston show that three white men and 32 colored men had died of pneumonia, three colored men of tuberculosis, four colored men of heart disease, one colored man of typhoid and five colored men from unclassified diseases, or 48 in all. In addition, three white men and 12 colored died of injuries, of whom 11 were in the tunnel. These accident figures are unusually low for projects of this size.

"Out of 5,000 adult males it is expected that some will die from time to time, as in any community. The vital statistics of Fayette County show that during the years the tunnel was being constructed the county annual death rate was 12 white males out of 1,000 and 23 colored out of 1,000. The death rate among our employees during this same period averaged six white males out of 1,000 and 24 colored out of 1,000. In the winter of 1931 pneumonia was prevalent throughout the county. It is interesting to note that the average deaths among our employees from pneumonia was considerably lower for the tunnel workers than for those employed outside.

"When suits were brought against this company by alleged victims of silicosis they were tried in the course of law and juries made up of residents of the neighborhood were unable to reach an agreement. In spit of this, we settled the claims against us, paying out large sums of money. Our attitude has been uncompromisingly fair. We did everything within our power to make the working conditions of our employees free from hazard, and we saw to it that any who died while in our employ received a decent burial.

"We brand as utterly false any contrary reports recently emanating from Gauley Bridge, W. Va."

THE SILICOSIS "RACKET"
Montgomery News, January 31, 1936

Right Up Rush's Alley

West Virginia's bottle-fed Senator Rush Dew Holt now has something that suits him exactly. It is playing up the racket end of the Hawk's Nest silicosis cases, which West Virginians heard about daily in the newspapers for many, many months several years ago and which they had every reason to believe had been settled, with everything ventilated about that matter that was to be ventilated.

We do not use the word "racket" without some authority. Dr. Harless, Gauley Bridge physician who was in close touch with the Hawk's Nest construction and who in pursuit of his medical practice had cause to study the silicosis situation and its effect upon workmen, has just told a Congressional investigating committee that suits for silicosis damage to health became a racket around Gauley Bridge.

A New York Congressman of foreign name and probable Bolshevist sympathies—Vito Marcantonio—appears to have re- stirred the silicosis matter for no other purpose than to gain support for legislation of a distinct Moscow pattern in Congress. His charges are that the laborers died like flies at Hawk's Nest, though official records are said to show a total of 48 deaths, very few of them remotely attributable to the disease known as silicosis. The New York Congressman challenges the record, and of course, Senator Holt is bold to say that official records are worthless and that there was "political" covering up the facts. Senator Holt has not yet said how long he has known what he "knows," or why, if he has known it for some time, he didn't sooner bring out the "tragic facts."

So long as the limelight lasts, it doesn't matter what the facts are to Mr. Holt. It cannot be imagined by those of us so close to the situation geographically as we are, that anything worse than the record could have been covered up so long and with such legal struggles as occurred over many months in the Fayette courts. We know that a big construction works was carried on, and full news details were published concerning several accidents which took lives, as will inevitably happen on such projects. Then we heard of the unexpected strike of silica rock or sand, at first regarded as a fortunate find, industrially speaking. Then came the long-drawn out litigation, hundreds of suits finally adjusted at a cost of several hundred thousand dollars to the contractors.

No one can object to a Congressional investigation which will seek that facts, whatever they may be, and is not designed for something else—or chiefly for cheap publicity—and which in the end may leave the country at large with an altogether erroneous impression of what occurred and the results upon human life.—***West Virginia News***

And His Number, Too

P.H. Faulconer, Virginia engineer and head of the contracting firm of Rinehart and Dennis, of Charlottesville, came back to Gauley Bridge to refute the lately-made sensational charges of wholesale silicosis deaths among his workmen and declared last Saturday that suits against his firm in 1931 were part of a 'silicosis racket." He asserted his firm had driven 26 tunnels in this and other states in 35 years, but had never even heard of the lung disease caused by breathing silica particles until 1931.

"These 450 suits filed against us in West Virginia," he continued, "are just a hold-up and blackmail. The silica content was about the same in the other tubes we drilled and I never heard of silicosis until this job was started."

Turning his attention to Senator Holt and the New York Representative who have said that hundreds died and 1,500 are doomed, Faulconer declared: "These statements are deliberate falsehood. Mr. Holt's statements are false. For sending a man to Washington who handles the truth so carelessly, the people of West Virginia should blush in shame."

Mr. Faulconer said that despite the fact that the Supreme Court of West Virginia had held that silicosis victims could not collect from the State's Workmen's Compensation Fund, that his firm had paid into the fund a total of "166,000, and that "since that time we have paid out over $300,000 to the silicosis racket." The records show that men brought suits who had never been in this tunnel, and some others by men who actually were in the tube an hour, according to our workmen's timesheets."

Perils of Rock Dust

The preamble of a bill introduced into the house by Representative Vito Marcantonio presents a harrowing picture of the ravages wrought by silicosis among laborers who drove the tunnel of the Hawks Nest hydro-electric project in West Virginia and of the cruelty of the contracting firm—too harrowing indeed for acceptance at its face value. Because of dry drilling and the lack of adequate ventilation it is charged that a locomotive headlight could not be seen five feet away, that foremen clubbed workers with pickhandles to make them enter the "tunnel of death," that 476 died of silicosis and that 1,500 more are suffering from the disease.

All this seems to be based on mountain gossip and an article published in a Midwestern radical weekly. What the facts are will no doubt be brought out by the investigation. Meanwhile the defense of the contractor, as the Engineering News-Record presents it, deserves as much publicity as the accusation. In the thirty months required to drive the tunnel 65 laborers died, according to his records—men who worked inside the tunnel and out. Accidents accounted for 15 of these deaths, manslaughter (the result of fights) for two, pneumonia for 35, tuberculosis for three, heart trouble for four, typhoid for one, and other unclassified causes for five. Considering that the number employed on the entire contract was 4,948 and that a total of 2,500 worked underground, the accident record is better than what engineers expect in tunnel work. The death rate is said to be less than half that of the county in which the tunnel lies. In view of a labor supply that greatly exceeded the demand the charge of clubbing is hardly credible. Ventilation is said to have conformed with the accepted standards. In addition to the usual first-aid organization, a physician was in residence.

Whatever the outcome of the investigation may be there can be no doubt that the perils of silicosis have been accentuated by the high speed and efficiency of the latest pneumatic tools and the fineness of the rock dust that they generate. Let minute cutting particles enter the lungs, and the end is death in five to twenty years. Such is the menace that the New York Department of Labor, when Miss Frances Perkins was at its head, prepared a code which calls for technical reforms and the granting of compensation to silicotic workers. The department even went so far as to sponsor an invention which catches rock dust at the drill hole—its source. If the investigation prompted by the reports from West Virginia is carried out objectively, if technical experts and physicians are called upon to suggest the proper method of dealing with one of the major hazards of excavation, Representative Marcantonio will deserve thanks. Silicosis is so dreadful a menace that more is wanted than a mere judgment of the case presented to his committee. Engineers would themselves like to know what standards should be observed to keep their tunnelers in good health and to protect themselves against charges of humanity.—***New York Times***

DOCTOR DENIES SO MANY DEATHS FROM SILICOSIS
Ohio Investigator Doubts if 200 Died from Breathing Silica Dust
Montgomery News, February 7, 1936

An authority on silicosis—an often fatal disease caused by breathing silica dust—disputed at Columbus, Ohio, Saturday charges that it caused 476 deaths on a New Kanawha Power Co. hydro-electric tunnel project at Hawks Nest, W. Va.

Dr. Emery R. Hayhurst, consultant in occupational diseases at the Ohio department of health, who has studied silicosis for 25 years, said he doubted more than 200 deaths were caused directly by the disease.

Rep. Vito Marcantonio of New York precipitated a congressional investigation by charging 476 workers on the project died form silicosis and that 1,500 more were dying.

Silica dust may have collected in the lungs and contributed to other deaths of pneumonia and tuberculosis, Dr. Hayhurst said.

He based his opinion largely on investigations made at Hawks Nest by a medical commission of which he was a member. The commission was hired by alleged silicosis sufferers who sued contractors on the project. Of 306 workmen who contended they were suffering with silicosis, Dr. Hayhurst said, only 152 were diseased.

The others had healthy lungs.

Some persons may be exposed to silica dust for 30 years, he said, before becoming afflicted with silicosis. Others may die of tuberculosis or pneumonia after being exposed only a few months.

The tuberculosis germ would have to be present in the lungs, he said, before silica dust was breathed, and it is a matter of dispute whether the dust aggravates tuberculosis or is antagonistic to it.

Symptoms of silicosis, the physician said, are shortage of breath, pains in the chest and poor circulation.

Silicosis easily can be discovered by the X-ray even when there are no outward symptoms.

Dr. Hayhurst recommended better control of the dust in mines and physical examinations of workers to cut down the toll of silicosis. Wet drilling and dust traps on drills, he said, will help much to eliminate the dust.

SILICOSIS NEW RACKET, CLAIM
Engineering and Mining Journal Would Probe Disease to End Racketeering
Montgomery News, March 20, 1936

Terming the public furore aroused in connection with silicosis deaths in a West Virginia tunnel the work of legal racketeers, the Engineering and Mining Journal, of New York, editorially urges in its forthcoming issue "Let's End the Silicosis Racket."

Behind it, declares the writer, is a record of legal racketeering in the form of damage suits for millions of dollars brought by enterprising lawyers on behalf of willing "victims"—all because of the "lack of adequate legislation on silicosis as a health hazard.

"Isn't it about time," it continues, "to take a realistic attitude toward silicosis and support a federal investigation of the whole subject by men of competence and authority."

GREAT TUNNEL AT HAWK'S NEST NOW GETTING WATER
Water Is Pouring Through Gates at Rate of Billion Gallons a Day

EXPECTED TUNNEL WILL BE FULL JULY
Will Stand About Two Weeks Before Being Turned Into Huge Turbines
Montgomery News, June 26, 1936

Water pouring through valves at the rate of more than a billion gallons a day is slowly filling the giant Hawk's Nest Tunnel on the New River.

Chief Engineer O.M. Jones, of the Electro-Metallurgical Company, who will use the 167-foot drop in the three and a half mile long boring to give the water sufficient velocity to turn four turbines capable of generating 140,000 horsepower of electricity, said at the present rate it will take until about July 1st to fill the tunnel. He added:

"Then we will let the water stand a week or so, to make sure everything is all right before turning it into the hydro-electrical plant."

A long trail of litigation, investigation and engineering accomplishment preceded the moment when workmen slowly opened the by-passes in the steel gate at the upper end.

More than 5,000 men labored for 30 months boring through long stretches of silica rock on a straight line through the West Virginia mountains to eliminate a huge S curse in the river. Before the $4,000,000 project, which included construction of a dam at the upper end near the famous Hawk's Nest rock which is a tourist's visiting point, the federal government stepped in with the contention that it has jurisdiction over the river. The supreme court of the United States finally turned down the suit on a technicality.

Later, a declaration by the State that the tiny particles of dust flying from drills biting into the silica rock are a contributing cause to various illnesses was augmented by charges in congress that 476 workmen had died and 1,500 others were stricken with silicosis, a lung malady. The West Virginia legislature made silicosis a compensable disease.

The labor committee of the house has a resolution by Representative Vito Marcantonio, Republican, of New York, to investigate the working conditions. A sub-

View into tunnel during construction, looking upstream, showing tunnel formwork preparatory to concreting. *Courtesy West Virginia State Archives.*

committee heard social workers and others who charged proper safety and ventilation measures were not taken by Rinehart and Dennis, Charlottesville, Va., contractors.

P.H. Faulconer, president of the firm, replied that the silicosis suits are "rackets," He added the company paid $170,000 to settle 300 cases out of court, and paid $166,000 to the state compensation fund although silicosis at that time did not come under the provisions of the law. He declared among 5.000 workmen, there were fewer deaths than there might be in a community of comparable size.

The state supreme court threw out 200 other suits.

Macantonio's charge that 169 workers were buried in a cornfield drew the reply from H.C. White, Summersville undertaker, that he had buried 27, all Negroes, in a private cemetery because there was none for Negroes in the community.

State officials declared the exact conditions probably never would be determined.

AFTERWORD

A NEW AND GREATER REGARD FOR HUMANITY?

"If by their suffering and death they will have made life safer in the future for the men who go beneath the earth to work, if they will have been able to establish a new and greater regard for human life, their suffering may not have been in vain."

WHEN THE CONGRESSIONAL SUBCOMMITTEE EXPRESSED THIS SENTIMENT in 1936, Frances Perkins directed the U. S. Department of Labor. Appointed by FDR in 1932 as the first woman to hold a cabinet level position, Perkins brought to the Labor Department an extensive technical background in the field of industrial hazards and hygiene. More importantly, however, was her near-boundless empathy and concern for the plight of the American worker.

Purportedly Perkins' response, when FDR offered her the position: "I don't want to say yes to you unless you know what I'd like to do and are willing to have me go ahead and try."

Her goals included providing direct federal aid to states for unemployment relief and public works; implementing a 40-hour workweek; initiating minimum wage laws; enacting child labor laws; and most importantly, ensuring safer environments for American workers. "Are you sure you want these things done? Because you don't want me for Secretary of Labor if you don't."

Perkin's pro-labor stance had little bearing on the silicosis problem at Hawks Nest, for a week after the silicosis hearings, a group of industrialists met privately at the Mellon Institute and established the Air Hygiene Foundation (AHF), "because of recent misleading publicity about silicosis …."[80]

"With the Air Hygiene Foundation, industry had found an effective propaganda formula: a combination of partial reforms with reassuring 'scientific' rhetoric, under the aegis of an organization with a benevolent, independent-sounding name."[81] Leading scientists and public officials were appointed as members and trustees of the foundation. Its spokesmen began to be widely quoted in popular trade publications. "'Silicotics are rare compared with men driven from their jobs by shyster lawyers,' commented AHF representative Alfred C. Hirth."

---

80. Air Hygiene Foundation: http://www.sourcewatch.org/index.php?title=Air_Hygiene_Foundation
81. Ibid., p. 1.

By 1940 the AHF had 225 member companies including representatives from American Smelting & Refining, Johns-Manville, United States Steel, Union Carbide and PPG Industries. In 1941 the AHF changed its name to the Industrial Hygiene Foundation (and later still to the Industrial Health Foundation, or IHF[82])....By the 1970s the Industrial Health Foundation had more than 400 corporate sponsors including Gulf Oil, Ford Motor Company, General Motors, Standard Oil of New Jersey, Kawecki Berylco Industries, Brush Beryllium, Consolidated Coal, Boeing, General Electric, General Mills, Goodyear, Western Electric, Owens-Corning Fiberglass, Mobil Oil, and Dow Chemical.

Relative to the issue of silicosis, Perkins convened the Tri-State Conference on Silicosis in Joplin, Missouri, April 23, 1940, with the following comments:

> *We in the Department of Labor have a mandate which we received from the Congress of the United States "... to foster, promote, and develop the welfare of the wage earners of the United States of America, to improve their working conditions, and to advance their opportunities for profitable employment."*
>
> *All of us here today have a social responsibility, a moral responsibility, and perhaps that lesser degree of responsibility which I call legal responsibility. For the latter is not as important as the social and moral responsibility....*
>
> *The owners and operators of these properties have a great moral responsibility.... The presence in this audience of a number of the ministers of religion reminds us, too, of the duty to bring about that kingdom of heaven on earth to which we all give lip service in order to bring about better conditions for all the people of the United States. [For] each individual, no matter how poor, still has tremendous worth and value; and in which the disaster of one is the concern of all.*[83]

As for Perkins' lofty sentiments, "...to foster, promote, and develop the welfare of the wage earners," one can easily imagine her reaction today to Section 2a of the 1970 Occupational Safety and Health Act which states that because "...personal injuries hinder interstate commerce; . . . in order to promote commerce, the workers must be protected;" obviously begging the question, "If personal injuries did not hinder interstate commerce, what priority would 'protecting the worker' assume?"[84]

In 2000 OSHA set forth remedial measures for preventing silicosis that are virtually the same preventative procedures well-known in the 30s—i.e. employing water to suppress dust; providing adequate fresh air and ventilation; and utilizing respiration filtration devices. Even so, and as relatively simple and inexpensive as these measures are, they were not employed at Hawks Nest; nor, apparently, in the workplaces of the 16,000[85] Americans who died from exposure to silica dust between 1968 and 2002.

---

82. Ibid., pp. 1-2.
83. http://historymatters.gmu.edu/d/128/
84. http://www.osha.gov/pls/oshaweb/owadisp.show_document?p_table=OSHACT&p_id=3356
85. Journal of Amaerican Medical Association. "Silicosis Mortality, Prevention, and Control," vol. 293, no. 21, June 1, 2005.

As for Union Carbide Corporation, its history speaks for itself. A quick review of the "Toxic Release Inventory" archived on the Environmental Protection Agency's website—www.epa.gov—reveals a pattern of spills and accidents that culminated with the grossly understated "unfortunate gas release" in Bhopal, India in 1984 that killed between 20,000 and 30,000 people. Although Union Carbide eventually agreed to a watered-down settlement with the Supreme Court of India, the corporation consistently refused to acknowledge responsibility for "the release." Dow Chemical purchased Union Carbide in 2001.

Albeit on the surface the incident at Hawks Nests appears linked to a specific era and a particular locale, in reality its message transcends all limitations of time and place; and likewise, in many respects, "…the lesson it suggests has yet to be learned: that it is not lack of knowledge which perpetuates occupational problems but a lack of commitment to change on the part of those with the power to do so.… Until workers' lives are considered more valuable than employers' profits, such tragic deaths will undoubtedly continue."[86]

Still today Hawks Nest remains a spot of unparalleled beauty, continuing to attract sight-seers from every corner of the earth; and frequently, if the timing's just right, visitors will partake of a unique serenade—the song of the siren adrift on the wind—reminding us once more—and again and again and again—(with apologies to John Donne,) to "ask not for whom the siren tolls—it tolls for thee."

---

86. Pat Forman, reprinted from Health/Pac Bulletin, November/December, 1977. *Mountain Life and Work*, August 1978. p. 32.

APPENDICES

A: BIOGRAPHICAL INFORMATION

VITO MARCANTONIO (1902-1954)

Vito Marcantonion, the son of Italian immigrants, represented an ethnically diverse congressional district in New York City's 20th district of East Harlem. As a congressman he was renowned for voting with his conscience rather than with his party—and was once described as "the loneliest man in Congress."

Constantly advocating social and labor reform, he attracted the scrutiny of the FBI and was subsequently investigated by the House Committee on Un-American Activities. During his fourteen-year tenure in Congress, he pushed for civil rights legislation long before the Civil Rights Movement, as we know it, came into existence.

Marcantonio lived his entire life in the Italian section of Harlem where he shared a modest brownstone with his mother and grandmother, adhering to the customs and culture of his local community. An anecdotal remark from that time period mentioned his impeccable "old-world style" manners—adding he was careful never to remove his jacket in the presence of women.

Upon his sudden death in August 1954, an estimated twenty thousand people viewed his funeral pier; and on the day of his funeral, ten thousand people wept openly in the streets of New York. The sense that swept through the crowd was, "Who will fight for us now?"

Marcantonio is buried in the Woodlawn Cemetery in the Bronx. The inscription on his tombstone:

VITO MARCANTONIO
DEFENDER OF HUMAN RIGHTS

MURIEL RUKEYSER (1913-1980)

"The universe is made of stories, not of atoms."
—Muriel Rukeyser

Born in New York City in December, 1913, Muriel Rukeyser was raised in an upper middle class family of Lithuanian-Jewish descent. The eldest of two daughters, she attended the Ethical Culture School in New York City, as well as Vassar College and Columbia University.

In 1933, at the age of nineteen, she attended the trial of the Scottsboro Nine in Alabama, and covered the antifascist Popular Olympics in Barcelona in 1936 as correspondent for London's *Life and Letters Today*.

Prior to the publication of *The Book of the Dead*, which appeared in her 1938 book *U.S. 1*, Rukeyser and a friend, Nancy Naumberg visited Gauley Bridge, West Virginia to gather information regarding the Hawks Nest Tunnel disaster. The following letter from Naumberg is located in the Muriel Rukeyser papers in the Library of Congress:

> *April 6, 1937*
> *Dear Muriel,*
> *I wanted to give you a few of my personal reactions to Gauley Bridge, and also to suggest a general outline. First, following your first two sentences, I would suggest describing the disease, and its symptoms. Then telling the story of Viv. Miller as we drove to view the tunnel, about which I had heard so much. Through his story, the background of the tragedy. It other words, as in the story I told Eliz. Do it chronologically, only this time getting in the facts as much as possible.*
>
> *Stress through the stories of Blankenship, Miller, etc. the necessity of a thorough investigation in order to indict the Co., its lawyers and doctors and undertaker, how the company cheated these men out of their lives, and the miserable conditions under which they now live; stress the relief situation, the inadequacy of it, how far they have to go to get it, how the silicosis men are put on the heaviest kind of work relief with the tunnel bosses in charge and how many of them are too sick to work, how when Jones and Robison testified, they were taken off work relief, and only put back on because of Congressional pressure.*
>
> *Stress the importance of silica rock—use and Robison's testimony for silica dust stories, show how we heard that the men working there have been bought off by the Co. Show how the tunnel itself is a splendid thing to look at, but a terrible thing to contemplate. Show how a similar condition must not be repeated, how there must be adequate precautions taken in industry, how adequate compensation laws must be enacted, how the whole thing is a terrible indictment of capitalism.*
>
> *Nancy*

An underlying theme of humanitarianism coupled with a strong commitment to social justice, permeates the entirety of Rukeyser's poetry—works for which she received the Copernicus, Shelley Memorial and Guggenheim Awards.

In 1943, challenging her patriotism on the basis of these poetic themes, the House Committee on Un-American Activities (HUAC) convened an investigation into Rukeyser's activities, concluding she was a member of numerous groups cited as communist front organizations, including the "Third American Writer's Congress," an organization that focused on the themes of "defense of democracy" and cooperation between nations.

In later years she taught at Vassar, Columbia University, Sarah Lawrence and the California Labor School. Alice Walker attributes her own literary success, in part, to Rukeyser's influence as both teacher and friend.

Rukeyser traveled to Hanoi in 1972 as a peace ambassador, and in 1975 stood in silent protest outside the jail cell of South Korean political prisoner and poet Kim Chi-Ha. She died in New York City at the age of 66.

ALBERT MALTZ (1908-1985)

Albert Maltz was born in Brooklyn and attended Columbia University and the Yale School of Drama. In 1936 his short story "Man on a Road," published in *New Masses*, influenced the Congressional Committee on Labor's decision to investigate the disaster at Hawks Nest.

Maltz' career as a screenwriter began in 1941, and in 1942 and 1945 his screenplays received academy award recognition. In 1947 the House Committee on Un-American Activities targeted Maltz and nine of his colleagues. Known as the "Hollywood Ten," this talented group of writers and producers challenged the right of HUAC to inquire into their political or religious beliefs. Excerpts from that testimony follow:

> *I am an American and I believe there is no more proud word in the vocabulary of man. As with any other writer, what I have written has come from the total fabric of my life—my birth in this land, our schools, our atmosphere of freedom, our tradition on inquiry, criticism, discussion, tolerance. Whatever I am, America has made me. And I, in turn, possess no loyalty as great as the one I have to this land, to the economic and social welfare of its people, to the perpetuation and development of its democratic way of life.*
>
> *Now at the age of 39, I am commanded to appear before the House Committee on Un-American Activities. For a full week this committee has encouraged an assortment of well-rehearsed witnesses to testify that I and others are subversive and un-American. It has refused us the opportunity that any pickpocket receives*

in a magistrate's court—the right to cross examine these witnesses, to refute their testimony, to reveal their motives, their history, and who exactly they are. Furthermore it grants these witnesses congressional immunity so that we may not sue them for libel for their slanders.

I maintain that this is an evil and vicious procedure; that it is legally unjust and morally indecent—and that it places in danger every other American, since if the right of any one citizen can be invaded, then the constitutional guaranties of every other American have been subverted and no one is any longer protected from official tyranny.

A few years ago, in the course of one of the hearings of this committee, Congressman J. Parnell Thomas said, and I quote from the official transcript:

I just want to say this now, that it seems that the New Deal is working along hand in glove with the Communist Party. The New Deal is either for the Communist Party or it is playing into the hands of the Communist Party.

Here is the reason I and others have been commanded to appear before this committee. In common with many Americans, I supported the New Deal. In common with many Americans I supported—-against Mr. Thomas and Mr. Rankin—-the anti-lynching bill. I opposed them in my support of OPA controls and emergency veterans housing and a fair employment practices law. I signed petitions for these measures, joined organizations that advocated them, contributed money, sometimes spoke from public platforms, and I will continue to do so.

I will take my philosophy from Thomas Payne, Thomas Jefferson, Abraham Lincoln, and I will not be dictated to or intimidated by men to whom the Ku Klux Klan, as a matter of committee record, is an acceptable American institution.

I would rather die than be a shabby American, groveling before men who names are Thomas and Rankin, and who now carry out activities in America like those carried out in Germany by Goebbels and Himmler.

Maltz was eventually convicted and spent nine months in a federal prison in West Virginia. Upon release he relocated to Mexico and continued working as both an advocate for personal freedom and as a writer. In 1969 the film industry recognized Maltz for the publication of his final screenplay, *Two Mules For Sister Sara*.

In 1978 together with a group of colleagues including Noam Chomsky, Philip Berrigan, Allen Ginsberg and Kate Millet, Maltz issued a letter to the Human Rights Conference advocating "human rights for everybody," and condemning the imprisonment of the Helsinki Watch Group.

Albert Maltz died in Los Angeles in 1985.

RUSH DEW HOLT (1905-1955)

Rush Holt was elected to the U.S. Senate in 1934 as a representative from the State of West Virginia. At the time of his election, he was twenty-nine didn't turn thirty until June, nearly five months after the swearing-in ceremony of January 3, 1935—a situation that created a great deal of debate locally and nationally.

During the Congressional investigation on silicosis, Holt advocated social reform—both in the work place and within society as a whole. "Soon after his accession to the Senate, Holt underwent a remarkable political metamorphosis. Believing his constituents were denied a fair share of the patronage emanating from federal relief programs, he began in 1936 by attacking the Works Progress Administration for political corruption and inefficiency.

Within months Holt emerged as one of the New Deal's most vocal conservative critics. Proudly independent, he sacrificed his alliance with Franklin Roosevelt, Matthew Neely, the United Mine Workers, and most rank and file Democrats in West Virginia. This maverick and impolitic behavior condemned Holt to a single Senate term. In the primary election of 1940 Holt placed third in his bid for re-nomination. Although he remained politically active and later joined the Republican Party, Holt failed to win another high office. Holt died from cancer in 1955 at age 49.

HOLT WILL FACE FIGHT IF HE APPEARS JANUARY 3
Fayette Tribune, 12/30/1934

Rush D. Holt, elected as United States Senator from West Virginia on the democratic ticket last November, will face a fight for his seat if he tries to take the oath of office January 3, when Congress convenes.

Senator McNary, of Oregon, has indicated that if the 29 year old Weston man presents his credentials, he will ask that they be referred to the elections committee.

Meanwhile, Holt's certificate of election has been forwarded to Washington from the West Virginia capital.

DECISION ON STATUS OF SEN. RUSH D. HOLT TODAY
Fayette Tribune, June 20, 1935

Today in the United States senate, in Washington, Rush D. Holt, democratic senator-elect from the state, will present his credentials for membership, and ask to be given the oath. He was 30 years old yesterday but party leaders tabled his resolution for one day by mutual agreement in order to take up the security bill. The ultimate decision will no doubt be momentous since no other exact precedent exists.

Although the minority party in the Senate has voiced opposition to his seating upon constitutional grounds, the elections committee, in parley early this week, voted 10 to 4 in favor of seating the young West Virginian. Their report will be considered in the debate today which forebodes much argument before he is finally voted on, because four senators representing the dissenting committeemen have indicated they will voice opposition on the floor. Although some conservative democratic solons are also opposed to Holt, his foes have conceded eventual defeat because of the large party majority in his favor.

Holt's supporters plan to submit supposedly identical cases to be used as criteria for his seating. His opponents, mainly republicans, are acting as defenders of the constitution. They believe that Holt is ineligible because he had not attained the age of 30 years at the start of his term. West Virginia has been without the services of two senators for six months.

ON SEATING RUSH HOLT—An Editorial:
Fayette Tribune, June 27, 1935

Our United States Senate took upon itself last week the responsibility of ignoring the spirit at least, if not the letter, of our Constitution, and of setting a precedent which can in the future, provide that a state may have no representation in that body, instead of the two representatives as provided by the Constitution.

In seating Rush D. Holt, the Senate declared in effect that it is n ot necessary to have attained the age of 30 before being elected or before drawing the salary and enjoying the rights and privileges of a senator, with the exception of actual participation.

The Senate said in effect, that a man or woman may be chosen by that body at the age of 24, and may, for a period of five years and 364 days, draw a salary of $10,000 a year, becoming a senator in fact on the last day of his "term" without having once represented his state in any particular.

And in making it possible for one man or woman to do that, it made it possible for any number to do so, thus, in theory at least, providing that a state or a number of states may at some future time, be without any representation whatever in that august—or rather now, ornate—body.

If the seating of Rush Holt were the only question, it would mean nothing. Another senator, more or less, in the United States Senate today means no more than another pebble, more or less, on the beach at Atlantic City. The problem of the individual in politics ceases to exist when his or her vote is not needed. And with a majority such as the Democratic part has in the Senate, surely none can say that Rush Holt will by his vote makr or break any measure. Therefore, whether he be a senator or not can mean but little.

But Rush Holt is not the question. Our form of government is at stake; whether we shall retain our Constitution is the problem.

It is not a party fight. That was proved when a number of Republicans voted to ignore the Constitution in this instance. It is a battle between those who believe in a democracy as planned by our forefathers, and those who believe something else should take it place. It is a fight between those who adhere to the principles of Washington and Jefferson, and those who would follow a rules, a dictator, as sheep will follow the bell weather, blindly, unknowingly.

We plead for Constitutional government. Amend that Constitution if you will; change it as provided by it. We sincerely believe that unless the people of this country insist upon adherence to it and its provisions, they will find themselves living under a despotism, a dictatorship, and we can't think they really want that.

Rush Holt should not have been seated. Not because he is Rush Holt; not because he is a Democrat; not merely because he was not 30 years of age. But because—and for this reason only—he was not eligible, and to declare him so was directly in conflict with our Constitution.

The victory was won by anti-Constitutionalists. It is high time the Constitutionalists asserted their mind.

HOMER "ROCKY" ADAMS HOLT (1898-1975)

A native of Lewisburg, WV, Homer "Rocky" Adams Holt received a degree in law from Washington and Lee University where, upon graduation, he served as professor of law from 1923 to 1925. Upon leaving W & L Holt moved to Fayetteville, WV, where he served as Chairman of the County Democratic Party and practiced law from 1925-1933, first with Hubbard, Bacon and Holt and then with Dillon, Mahan, and Holt. This latter firm represented the defendants Rinehart and Dennis and the New Kanawha Power Company during the silicosis trials in Fayetteville.

In 1932 Holt was elected attorney general of West Virginia, a position he held until 1936 when he was elected governor of the state. During his term as attorney general, Holt successfully defended Union Carbide in the Supreme Court trial, UNITED STATES V. STATE OF WEST VIRGINIA, 295 U.S. 463 (1935).

While governor from 1937 to 1941, Holt successfully blocked publication of the WPA Writer's Project book, *West Virginia: A Guide to the Mountain State* largely due to the unfavorable portrayal of Union Carbide's complicity in the Hawks Nest Tunnel disaster.

Holt practiced law with the firm of Brown, Jackson and Knight in Charleston from 1942-1946; and in 1947 assumed the position of general counsel to the Union Carbide and Carbon Corporation in New York where he also served as a director and vice president until 1953.

After retiring in 1953, he practiced law in Charleston with the firm Jackson, Kelly, Holt and O'Farrell until his death in 1975.

B: RELATED DOCUMENTS

1. Electro Metallurgical Company Hawks Nest Hydro Electric Development: Exhibit B. Land tracts within the "project property line." 194

2. New Kanawha Power Company, "Notice to Form Corporation," 1927. 196

3. WV Public Service Commission: New Kanawha Power Companies' notice to transfer all rights and property to the Electro Metallurgical Company, April 1933. 197

4. New Kanawha Power Company, dissolution of corporation, 1935. 198

5. Partial Testimony of Dr. Emery R. Hayhurst. 199

6. Sections of contract between Rinehart & Dennis and the New Kanawha Power Company. 203

7. Hawks Nest Project: Examples of Bid Items. 204

8. "Statement of Cost" for construction of Hawks Nest project, September 30, 1936. 205

9. Defendant's instructions to the jurors, in the silicosis trial of "Raymond Johnson v. Rinehart & Dennis, et. al." 207

10. Miscellaneous correspondence prior to construction of the tunnel:
    Telegram, J.T. Hatfield to PSC, August 20, 1928. 209
    Telegram, PSC to J.T. Hatfield, August 21, 1928. 209
    Chesapeake and Ohio Railway Company to PSC, September 22, 1928. 210
    Great Kanawha Valley Improvement Association to PSC, October 13, 1928. 211
    J.T. Hatfield to A.B. Moore, Secretary of PSC, October 23, 1928. 211

11. Memorandum of the West Virginia PSC to the Federal Power Commission, re: "Matter of Declaration," January, 1929. 212

12. Request to "Modify Plans and Build Only One Dam and One Power Plant," July, 1929. 214

13. Letter from the Campbell's Creek Coal Company, September, 1929. 214

14. J.E. Settle reports, 1930-1933. 216

15. Replacement of J.E. Settle as PSC's inspecting engineer. August, 1933. 222

16. U. S. Supreme Court ruling, UNITED STATES V. STATE OF WEST VIRGINIA, 295 U.S. 463 (1935). 223

17. Miscellaneous Correspondance:
    New-Kanawha Power Company Tax Department to PSC, June, 1932. 229
    PSC to Clyde B. Johnson, June, 1932. 230
    PSC to F.M. Livesey, June, 1932. 231
    Benjamin G. Smith to Clyde B. Johnson, June, 1932. 232
    Clyde B. Johnson to Wade Coffman, August, 1932. 235
    FPC to New Kanawha Power Company, December, 1933. 237
    FPC Declaration of intention, January, 1934. 238
    Telegram, FPC to West Virginia PSC, December, 1934. 240
    Telegram, FPC to WV PSC, December, 1934. 240
    Letters from Dr. Emery Hayhurst to Dr. L.R. Harless, 1936. 241

18. PSC Ruling re: Case No. 1863, "Electro Metallurgical Company, Successor and Assignee of New Kanawha Power Company," December 1956. 243

19. PSC Ruling re: Case No. 1864, "Electro Metallurgical Company, Successor and Assignee of New Kanawha Power Company," December 1956. 244

20. Computation of Annual Charge for the Hawks Nest Hydro-Electric Project of New River in West Virginia. 245

ELECTRO METALLURGICAL COMPANY
HAWKS NEST HYDRO ELECTRIC DEVELOPMENT[87]
West Virginia

EXHIBIT B

LAND, WATER RIGHTS AND TUNNEL RIGHTS LYING WITHIN THE PROJECT PROPERTY LINE

Public Service Commission Account No. 311

SUMMARY OF VALUES

Tract Numbers listed below refer to tracts as shown on the "Hawks Nest Hydro Electric Power Development Property Map" and only such portion of each tract as lies within the "Project Property Line" shown on that map is included in the cost as itemized.

LEFT RIVER BANK

Fayetteville District
    Cataract Land Co.   Tract Nos. 9 to 19 $21,000.00
    Eli J. Taylor   Tract No. 8   78.03
    Annie L. Parks   Tract Nos. 4 and 5   1,187.65

Kanawha District
    J.F. Meagher   Tract No. 19   1,800.00
    J.F. Meagher   Tract No. 20   1,300.00
    J.T. Morehead   Tract No. 18   500.00
    T.A. Dietz   Tract No. 27   1,000.00

TOTAL – LEFT RIVER BANK $26,865.68

RIGHT RIVER BANK

Mountain Cove District
    National Water & Power Co. $38,459.20

---

87. Public Service Commission. 75-PSC-21. Case 1863-1864. MicroFile #436B.

Franklin Real Estate Co.	Tract No. 42	1,100.00
Brockman Smokeless Coal Co.	Tract No. 53	5,901.25
W.A. Ohley, Trustee	Tract No. 52	11,692.25
D.D. Russell & Geo. Love, Trustee	Tract Nos. 64, 82, 83 & 84	12,115.07
D.D. Russell	Tract Nos. 42 and 43	8,363.69
Kate Young Dempsey	Tract Nos. 44 to 47)	11,417.00
R.T. Hubbard	Tract Nos. 48 to 51)	
Gauley Mt. Coal Co.	Tract Nos. 40 and 41	1,769.25
A.W. Hamilton	Tract Nos. 26 to 37)	39,639.72
A.W. Hamilton	Tract Nos. 55 to 60)	
Laura B. Taylor	Tract Nos. 38 and 39	2,108.44
W.F. Grady	Tract No. 22	910.37
Virginia F. Downey	Tract No. 20	3,237.88
Virginia F. Downey	Tract No. 21	1,409.18
Virginia F. Downey	Tract No. 23	523.07
Azel Meadows, Jr.	Tract No. 19	442.18
W.H. McVey	Tract No. 16	1,786.74
W.H. McVey	Tract No. 17	312.99
W.H. McVey	Tract No. 18	110.00
Azel Meadows, Jr.	Tract Nos. 14 and 15	2,761.28

RIGHT RIVER BANK (CONTINUED)

Mountain Cove District (continued)

L.B. O'Neal	Tract Nos. 10 to 13	2,003.75
J.V.R. Skinner	Tract No. 8)	3,806.00
W.H. McVey	Tract No. 9)	
C.W. Carver	Tract Nos. 6 and 7	3,712.47
E.M. Dalporto & S.A. McGraw	Tract Nos. 5 and 85	3,568.50

Falls District

C.T. Hawkins	Tract No. 148)	417.99
W.A. Hawkins	Tract No. 149)	
T.L. Kearse	Tract No. 150	4,940.00
Gauley Mt. Coal Co.	Tract No. 147	23,133.50
J.F. Meagher	Tract No. 146	2,556.50
E. Peyatt	Tract No. 145	2,113.87
Z.V. Cales	Tract No. 143	1,859.57
J.T. Morehead	Tract No. 144	2,000.00

TOTAL – RIGHT RIVER BANK		$193,676.98

J.M. Morehead	Tract Nos. 1, 24, 25 and 63 Mt. Cove Dist.)	
	Tract No. 151, Falls District)	
	Tract Nos. 1 to 3, 6 and 7 Fayetteville District)	
	Tract No. 21, Kanawha District)	$97,991.00

LAND SURVEY CHARGES	$11.637.07

TOTAL ACOCUNT NO. 311 LAND AND FLOWAGE RIGHTS	$330,170.73

NOTICE TO FORM CORPORATION

The undersigned agree to become a corporation by the corporate name of NEW-KANAWHA POWER COMPANY.

...The principal office and place of business of this corporation shall be located in the Town of Glen Ferris, in the County of Fayette and State of West Virginia. Its chief works will be located in the Districts of Falls and Kanawha and County of Fayette and elsewhere in the Sate of West Virginia.

...With objects and purposes to generate, produce, sell and distribute hydraulic, electrical and/or other power produced by water, for any other purpose whatsoever, both public and private, and to carry on the business of supplying and selling power.

...And to construct, purchase...or in any manner acquire dams, reservoirs, canals, aqueducts, tunnels, conduits and other waterways whereby water from the New River and the Kanawha River and/or other rivers in the State of West Virginia...will be utilized as necessary for the conduct or convenience of its business.

Given under our hands this 7th day of January, A.D. 1927.
Leonard Davis
Fred F. Haggerson
Edward W. Burdick
Benjamin G. Smith
Edward S. Whitney

Carbide and Carbon Building,
30 East 42nd Street, New York, New York

THE PUBLIC SERVICE COMMISSION
OF WEST VIRGINIA
CHARLESTON[88]

At a session of THE PUBLIC SERVICE COMMISSION OF WEST VIRGINIA at the Capitol in the City of Charleston on the 24th day of April, A. D., 1933.

CASE NO. 1864

NEW-KANAWHA POWER COMPANY,

Application for authority to construct dam, hydro-electric plant and tunnel on New River in Fayette County, West Virginia. (Hawks Nest Dam.)

The 24th day of April, 1933, came New Kanawha Power Company by Clyde Beecher Johnson, Esq., of its counsel, and presented for record a paper writing made by said New Kanawha Power Company and Electro Metallurgical Company <u>transferring the rights and property involved in this proceeding to said Electro Metallurgical Company</u>, which paper writing is noted of record in the words and figures following:

KNOW ALL MEN BY THESE PRESENTS, That New Kanawha Power Company, a corporation organized and existing under the laws of the State of West Virginia (hereinafter called the "Transferor"), for and in consideration of the sum of One Dollar ($1.00) hereby agreed to be paid to it by ELECTRO METALLURGICAL COMPANY, a corporation organized and existing under the laws of the State of West Virginia (hereinafter called the "Transferee"), and of other good and valuable considerations, has sold, assigned, transferred and set over and by these presents does sell, assign, transfer and set over unto the Transferee, its successors and assigns, under and by virtue of, upon the conditions provided in, and to the full extent authorized by, chapter one hundred fifteen, acts of the legislature of West Virginia, one thousand nine hundred thirty-three, those certain permits, together with the rights, powers and authority appurtenant thereto or in anywise thereunto belonging, granted to the Transferor by the public service commission of West Virginia under and by virtue of chapter seventeen, acts of the legislature of West Virginia, one thousand nine hundred fifteen, and as set forth in orders of the said public service commission made and entered by it in certain cases identified by it in its official records as case number one thousand eight hundred sixty-three and case number one thousand eight hundred sixty-four, and consolidated as one project and identified as said case number one thousand eight hundred sixty-four, and popularly known as the Hawks Nest-Gauley Junction project; and the transferee does by these presents accept the said sale, assignment, transfer and setting over of the said permits, together with the said rights, powers and authority appurtenant thereto or in anywise thereunto belonging, amended in the manner provided for in chapter one

---

88. Ibid.

hundred fifteen, acts of the legislature of West Virginia, one thousand nine hundred thirty-three, and upon the conditions upon which the sale and transfer of the said permits, together with all of the rights, powers and authority appurtenant thereto, are authorized by chapter one hundred fifteen, acts of the legislature of West Virginia, one thousand nine hundred thirty-three.

IN WITNESS WHEREOF, the Transferor and the Transferee have caused this instrument to be executed, in triplicate, by their respective corporate officers thereunto duly authorized and their respective corporate seals to be hereunto affixed, this nineteenth day of April, 1933.

STATE OF WEST VIRGINIA
CERTIFICATE[89]

I, WM. S. O'BRIEN, Secretary of state of the State of West Virginia hereby certify that

F. F. Haggerson, President of
NEW-KANAWHA POWER COMPANY,

A corporation created and organized under the laws of the State of West Virginia, has certified to me under his signature and the corporate seal of said corporation, that, at a meeting of the stockholders of said corporation, regularly held in accordance with the requirements of the law of said State, at the office of said corporation in Glen Ferris, West Virginia, on the 29th day of May, 1935, at which meeting more than sixty percent of the issued and outstanding voting stock of each corporation being represented by the holdrs thereof, in person, by bodies corporate or by proxy and voting for the following resolution, the same was duly and regularly adopted and passed to wit:

"RESOLVED, That the NEW-KANAWHA POWER COMPANY, a corporation created and organized under the laws of the State of West Virginia, does hereby discontinue business as a corporation and surrenders to the State its charter and corporate franchises.

FURTHER RESOLVED, That the president or vice-president of this corporation is authorized and directed to certify these resolutions to the Secretary of State of West Virginia and cause notice of the adoption of the foregoing resolution of dissolution to be published once a week for two successive weeks in some newspaper of general circulation, published in the county in which is located the principal office or place of business of this corporation.

---

89. Ibid.

And I further certify that a proper certificate of the publication of the notice required by section 80 of article 1, chapter 31 of the Code of West Virginia by the said corporation has been filed in my office.

WHEREFORE, I do declare that said notice was duly published, and that the dissolution of said corporation as set forth in the foregoing resolution is authorized by law.

Given under my hand and the Great Seal of the said State,
    At the City of Charleston, this <u>Twenty-sixth</u>
    Day of June, 1935.
    WM.S. O'BRIEN
    Secretary of State

PARTIAL TESTIMONY OF DR. EMERY R. HAYHURST[91]
In the Trial of Raymond Johnson vs. Rinehart & Dennis, Inc.

Direct examination by Mr. Bacon:
Q. Have you been sworn, Doctor?
A. Yes, sir.

[omission]

Q. What is your profession?
A. I am a physician and I specialize in occupational diseases.

[omission]

Q. What are you engaged in at the present time, Doctor?
A. I am consultant to the Ohio State Department of Health in occupational diseases, a position that I have held for 20 years this May. I am a consulting hygienist to the United States Public Health Service, do a certain amount of work in the capacity also for the United States Bureau of Mines, and indirectly for the United States Bureau of Standards.

[omission]

Q. Is silicosis an occupational disease?
A. It is.

---

90. "An investigation…." Pg. 82-95.

[omission]

Q. How long has it been known as such?
A. Well, it has been known as such almost since silica has been identified by the chemists. That is over a hundred years. But long before silica was identified, its effects were known. In the ancient days of the skilled trades the craftsmen—in fact, silicosis is probably the oldest occupational disease. There is reason to believe that it existed before written history, in connection with the making of arrowheads and spearheads, and so forth, by ancient warriors.
Q. How long has it been known and considered as an industrial hazard in the United States?
A. About 30 years. Twenty-five to 30 years. I should judge. It has been identified as such in the United States since 1915.
Q. Tell the jury what you mean by silicosis. Describe it, if you will.
A. Silicosis is a disease which results from silica dust in the air being inhaled into the lungs, where it sets up a gradual scar tissue formation in a peculiar way. It does not do it in a broad sheet like manner, but in spots, or nodules, as they are called, and those gradually increases in size until they block up the drainage from the lungs, and eventually they block the blood supply, and then they block the air passages. While this is going on the person suffers certain symptoms, such as increasing shortness of breath and can't take a deep breath. He gets pains around through his chest and he notices that he can't work as vigorously as before. He is very apt to develop tuberculosis, much more susceptible as the result of this silica in his lungs and otherwise—

[omission]

Q. What is it about the silica dust that makes it harmful and dangerous to human life when breathed?
A. Well the dust is not irritative immediately, so that the lungs and the respiratory passageways do not tend to throw it out as it comes in, which is true of most other dusts. As a result, it very subtly gets into the tissues of the lungs. Once in it begins to undergo a very slow dissolving, and when it gets in a dissolved state it is poisonous, and as the result of that poisoning nature, tries to wall ip up, and scar tissue begins to form around the particles of the dust, even one particle, a little mass of particles, and that scar tissue is what we see in the x-rays as these nodules all over the plates.

[omission]

Pg. 87
Q. Did anyone show you the lungs of Cecil Jones and also Walter Street?
A. Yes, sir.
Q. Who was that?
A. Dr. Harless.

Q. Did you make an examination of those sets of lungs?
A. I did.
Q. I wish you would tell the jury whether or not those lungs were silicotic.

[omission]

A. They were.
By Mr. Bacon:
Q. What evidence did you find in the lungs themselves of that condition?
A. The naked eye appearance of the lungs was enough to establish that diagnosis. They had black pigmented areas on the surface of the lungs almost any place you looked in the cross section of the lungs, and those areas and those areas were palpable—that is, they were able to be felt. They stuck up above the surfaces. They were, therefore, They were scar tissue. nodular. Ordinary coal dust pigmentation would not do that. You would have the black areas everywhere, but they would lie flat. You couldn't feel them. If your eyes were shut, you would not know they were there. But with these lungs, the outside feel of them, even after they had been in formalin, 10 percent, I believe he said—by sight and by feeling, that these lungs had scar-tissue nodulation. The only substance that we know that will produce that in a uniform manner—that is scattered all over the lungs—is silica dust. I further examined them microscopically.
Q. What did you observe as to the elasticity of these two lungs compared to the normal lung? Just describe that feature of it, will you?
A. Well, of course, I recognized that these lungs had been in formalin and that would reduce their elasticity. It would tend to stiffen and harden somewhat, as it does in all tissue preserved in that manner, but as compared with normal lungs you can compress them practically shut, like you do a wet sponge, and they spring out again to their normal size and shape, but these lungs you couldn't compress them; hardly any impression could you make. We expect some of that in any preserved tissue. Perhaps you can take an ordinary lung, preserved in formaldehyde, as I understand those were, maybe 6 or 8 months, and compress them half of the way or two-thirds of the way, but you couldn't compress these lungs one-fifth of the way.
Q. Tell the jury just why they could not be compressed?
A. They were so full of scar tissue nodules that you couldn't squeeze—you can't squeeze a rock together, that is the size of it.

[omission]

Pg. 90
Q. Doctor, I wish you would resume your original thought where you were giving the different conditions that influenced the time of exposure necessary to contract the disease.
A. Well, naturally, whether the dust is wet or dry is a great factor. Wet silica dust has been found to keep it out of the way to the extent of 90 to 95 percent. Even at that,

silicosis may follow; or say, 5 to 10 percent may b e enough under the conditions to product it. If it is not, a man breathes more deeply, and therefore inhales more silica dust.

[omission]

A. If the air is impure—that is , not rich enough in oxygen—the man breathes more deeply and naturally takes in more dust. If the air contains poisonous gases that then to displace part of the oxygen of the air, naturally the man has to breathe more to get the amount of oxygen he needs, so he takes in more dust.

[omission]

The witness. If the ventilation in the presence of gasoline exhaust fumes or motor exhaust fumes is so poor that you can smell it, it increases the hazard because one has to breathe deeper to get his oxygen. He has to breathe deeper to get his oxygen because of the presence in these gases of carbon dioxide in large quantities and carbon monoxide in smaller quantities

[omission]

Pg. 91
Q. Doctor, do you know Raymond Johnson, the plaintiff?
A. Yes, sir.

[omission]

Pg. 95
Q. What is the probably prognosis of Mr. Johnson, assuming there will be no complications set in?

[omission]

A. His prospects are bad for continued health, and even worse for continued longevity.

[remainder of text omitted]

Provisions of Contract[91]
between
Rinehart and Dennis and
Union Carbide, New-Kanawha Power Company:

On or before April 1, 1930, the Contractor shall have assembled and installed sufficient plant to enable it to make an effective beginning of, and shall actually begin, the construction of The Work. Thereafter The Work shall be prosecuted continuously and with such rapidity that within the period of twenty-three (23) months after the date hereof all the work to be done hereunder for the Dam and Power House shall be completed and within the period of two (2) years after the date hereof all The Work to be done under this Agreement shall be completed.

If in the opinion of the Engineer, the Contractor shall not have completed The Work to the said extent within the said period of two (2) years, then from the payment to the Contractor upon the Completion of The Work shall be deducted by the Company, as and for liquidated damages to the Company, the amount of $250.00 for each day by which the time of completion to said extent exceed the said period of two (2) years. If in the opinion of the Engineer, the Contractor shall have fully completed The Work within less than the said period of two (2) years, then to the payment to the Contractor upon the completion of The Work shall be added by the Company, as and for a premium to the Contractor, the amount of $250.00 for each day by which the time of full completion is less than the said period of two (2) years;

Access to the Work:
The Company and the Engineer, agents and employees of the Company may at all times enter upon The Work and premises used by the Contractor or into his works or shops, and the Contractor shall provide safe and proper facilities for such entrance and for the inspection of materials and workmanship.

Responsibility for the Work:
The Contractor shall take all responsibility of The Work, and take all precautions for preventing injuries to persons and property in or about The Work; shall bear all losses resulting to it on account of the amount or character of The Work, or because the nature of the land in or on which The Work is done is different from what was estimated or expected, or on account of the weather, elements or other cause.

The Contractor shall pay all debts for labor and materials contracted for by it on account of The Work herein contemplated. The Contractor shall assume the defense of, and indemnify and save harmless, the Company and its officers and agents, from all claims relating:

---

91. Ibid.

To labor and materials furnished for The Work;

To injuries to any person or corporation or to damages to any property of any person or corporation caused by the acts or negligence of the Contractor or any of its agents or employees, or in consequence of any improper materials, implements or labor used therein; and shall fully reimburse and repay to the Company all outlay and expense which the Company may incur by reason of its failure so to do. In any case where, in the opinion of the Engineer, injuries to any person or corporation or damages to any property are likely to result from any acts or negligence of the Contractor, or any of its agents or employees, the Engineer shall have the right to employ such measures as he may deem necessary or desirable to effect a satisfactory avoidance of such injuries or damages, and if, in his opinion, the case appears urgent, he may proceed to employ such measures without previous notice to the Contractor which, however, shall not be relieved from any responsibility on account of such action of the Engineer....

Examples of BID ITEMS[92]

When contractors bid on the entire Hawks Nest project—including excavation of the tunnel and construction of the dam and power house—each item of the task was broken down into very specific line items. The following are but a few of the delineated aspects of the entire project for which Rinehart and Dennis procured the contract:

Item No:	Description	Amount
2.	Earth Excavation	$1.50 per cu. yd.
3.	Rock Excavation	$4.00 per cu. yd.
4a.	Mass concrete, approximately 1:3:5 mixture, using local sandstone aggregates, not including cement	$6.50 per cu. yd.
4b.	Mass concrete in spillway section, approximately 1:2:4 mixture using	

---

92. Ibid.

	crushed local aggregates, not including cement	$6.50 per cu. yd.
6a.	Furnishing reinforcing steel in operating bridge (subject to substitution as per specifications)	$0.04 per pound
6b.	Placing reinforcing steel in operating bridge	$0.02 per pound
7.	Placing 9 ft. discharge valves and appurtenances in dam	$500.00 per valve
8a.	Steel racks at inlets to valves	$0.08 per pound
8b.	Drainage pipes and fittings in dam	$0.10 per pound

"Statement of Cost, September 30, 1936"[93]

Presented to the Public Service Commission on December 7, 1936:

P.S.C. Account #	Description	Cost
303	Miscellaneous Intangible Capital	$ 9,377.95
311	Land and Flowage Rights	330,170.73
312	Structures Power House Railway Spur and	$584,040,31

---

93. Ibid.

	Road to Power House	81,577.93	$665,618.24
318	Dam and Waterways		
	Dam		$1,392.506.91
	Accessories to Dam		
	Road to Dam	37,202.44	
	Change in C&O Bridge	25,338.39	
	Change in County Road	11,687.84	1,466,735.58
	Intake		175,027.96
	Tunnel and Penstocks		4,747,568.41
	Accessories to Tunnel		
	Surge Basin	217,920.77	
	Surge Tank	82,139.52	
	Tail Race	34,516.22	5,082,144.92
	Total of Accounts Nos. 312 and 318		7,389.526.70
Less Credit in Settlement with Contractor		114,246.90	
Net Total of Accounts Nos. 312 and 318			7,275,279.80
320	Water Turbines	527,983.70	
322	Electric Plant – Hydro Generators and Switching Equipment	781,491.63	
328	Sub-Station Equipment		
	Outdoor Sub-Station	139,006.53	
	Step-Up Transformers	225,866.23	364,872.76
331	Transmission Line		
332	From Power House to Dam		16,608.42
344	General Equipment		44,677.37
351	Engineering and Superintendence		1,124,785.78
		TOTAL	$10,475,248.14

DEFENDANT'S INSTRUCTIONS TO THE JURORS:

Note: The following instructions were located in the Fayette County Circuit Clerk's office, in the "Raymond Johnson v. Rinehart & Dennis, Inc., and E.J. Perkins" files.

Defendants' Instruction No. 2: (Given as Amended)

The Court instructs the jury that when the plaintiff, Raymond Johnson, accepted employment with the defendant, Rinehart & Dennis, and went to work in the tunnel, he assumed the risk of all dangers normally and ordinarily incident to his employment while working for the defendant whether he was actually aware of such dangers or not.

And the Court further instructs the jury that if they believe from the evidence that the injury complained of by the plaintiff was caused by his breathing dust from the rock through which the tunnel was being driven, and that the breathing of such dust was a danger normally and necessarily incident to the work in which he was so engaged, and could have been reasonably anticipated by him, then the jury should find for the defendants, even though the plaintiff may not have been aware of the extent of such danger.

Defendants' Instruction No. 3: (Given as Amended)

Paragraph 2: ... the Court further instructs the jury that if they believe from the evidence that the plaintiff, while working in said tunnel, discovered that it was dangerous to breathe the dust therein, and continued to work therein, and that he thereafter contracted the injury complained of by him due to his breathing of such dust, then the jury should find for the defendants.

Defendants' Instruction No. 7:

The Court instructs the jury that the mere happening of an injury is of itself no evidence of negligence. There must be affirmative and preponderating proof of negligence in the construction of the tunnel by the defendants showing more than a mere probability of such negligent construction before the plaintiff can recover, and the Court further instructs the jury that the burden of proof is on the plaintiff to prove by a preponderance of affirmative evidence the negligence charged in the declaration and if the jury believe that the evidence of any point necessary to maintain the plaintiff's case is evenly balanced or preponderates in favor of the defendants, then the jury shall find for the defendants.

Defendants' Instruction No. 8:

The Court instructs the jury that even though they may believe from the evidence that the defendants were negligent in conducting operations in the construction of the tunnel, nevertheless, if they further believe from the evidence that the plaintiff knew of the dangerous conditions existing in the tunnel, or as a reasonably prudent man should have known of them, and continued to work where he was exposed or might be exposed to breathing rock dust, then the fact that he continued so to work where he was, or might be, exposed to breathing rock dust constitutes contributory negligence on his part and bars a recovery by him, in which case the jury should find for the defendants.

Defendants' Instruction No. 9 (Given as Amended)

The Court instructs the jury that it is its duty if possible to reach a verdict which will accord with the law and the evidence in this case, but that in order to render a verdict, they must unanimously agree on such verdict; and that it is wrong for any juryman to allow any issue to be decided by a majority vote against his belief derived from the evidence; and that each juryman must be guided by his own belief from the evidence, even though such course may result in a failure of the jury as a whole to agree.

Defendants' Instruction No. 17-A (Given as Amended)

(Last paragraph) ..., if the jury believe that the defendants exercised the same degree of care as any other reasonably prudent person or company, knowing of such high silica content, but not knowing of the extra hazard or danger by reason thereof, and after having used reasonable care in the premises, had no reason to know thereof, would have exercised in the construction of the tunnel, then their verdict should be for the defendants.

Defendants' Instruction No. 19 (Given as Amended)

The Court further instructs the jury that if they believe from the evidence that during the time Raymond Johnson was employed by the defendant, Rinehart & Dennis, in the tunnel mentioned in the evidence, the defendants did not know about the disease called silicosis or similar disease arising from the inhalation of silica dust, and the said defendants were not negligent in not knowing about the said danger, then and in that event you should find for the defendants, even though you further believe from the evidence that the inhalation by the plaintiff of silica dust in said tunnel was harmful to the health of said plaintiff.

Defendants' Instruction No. 20

The Court instructs the jury that there is evidence in this case to show that the plaintiff is suffering either from silicosis or tuberculosis, or both diseases, and that under the law the defendants are not liable if he is suffering from tuberculosis and not from silicosis. Therefore, if the jury does not believe that he is suffering from both diseases and it is just as probable that he is suffering from one disease as the other, then the jury cannot guess which disease he is suffering from, and their verdict should be for the defendants.

WESTERN UNION[94]

NEWCOMB CARLTON, PRESIDENT J.C. WILLEVER, FIRST VICE-PRESIDENT

Received at 803 Kanawha Street, Charleston, W. Va. 1928 AUG 20 AM 11 20

HNA 106 61 DL=CB LAKE PLACID CLUB NY 20

THE PUBLIC SERVICE COMMISSION=
CHARLESTON WVIR=

JUST BEEN ADVISED BY WAR DEPARTMENT YOU ARE CONSIDERING PERMIT TO THE NEW KANAWHA POWER CO FOR HYDRO PROJECT HAWKS NEXT STOP IN BEHALF OF THE HATFIELD CAMPBELL CREEK COAL CO THE LARGEST SHIPPERS OF TONNAGE ON KANAWHA RIVER WE PROTEST EARNESTLY AGAINST GRANTING PERMIT AND BEG FURTHER CONSIDERATION UNTIL THE TRANSPORTATION INTEREST CAN MAKE PROPER EXAMINATION FOR APPROVAL OR DISAPPROVAL=
J T HATFIELD.

THE QUICKEST, SUREST AND SAFEST WAY TO SEND MONEY IS BY TELEGRAPH OR CABLE

WESTERN UNION[95]

NEWCOMB CARLTON, PRESIDENT J.C. WILLEVER, FIRST VICE-PRESIDENT

Charleston, West Virginia, August 21, 1928.

Col. J.T. Hatfield,
Lake Placid Club, New York.

Your telegram presenting against permits to New Kanawha Power Company behalf

Hatfield Campbell Creek Coal Company received stop

---

94. Ibid.
95. Ibid.

Applications of power company

And protest set for hearing September fifth stop Copy order follow by mail.

THE PUBLIC SERVICE COMMISSION

(charge to the Public Service Commission)

THE CHESAPEAKE AND OHIO RAILWAY COMPANY[96]
Huntington, W. Va., September 22nd, 1928

File 191/142

Case No. 1863 – New Kanawha Power Company, Application for authority to construct dam, hydro-electric plant and tunnel on New River in Falls District and Kanawha District, Fayette County, West Virginia. (Gauley Junction Dam.)

Case No. 1864 – New Kanawha Power Company, Application for authority to construct dam, hydro-electric plant and tunnel on New River in Mountain Cove District and Fayetteville District, Fayette County, West Virginia. (Hawks Nest Dam.)

Public Service Commission,
Charleston, W. Va.,
Gentlemen:

In the above matter, should you decide to grant these permits, or either of them, will you kindly, before finally granting or entering them, advise us of the contents?

We believe, that in order to properly protect the rights of the Chesapeake and Ohio Railway Company, the permits, or either of them if granted, should provide that the improvement should be constructed, maintained and operated under and in accordance with the terms and conditions of the deed and agreement of March 26th, 1924, between the Electro Metalurgical Company and The Chesapeake and Ohio Railway Company.

Will you kindly advise us direct or through your Secretary further in regard to this matter?

Yours very truly,
FITZPATRICK, BROWN & DAVIS
State Counsel
By: H.S. King

---

96. Ibid.

HSK
Copy to Mr. Taylor.

GEO. E. SUTHERLAND   C.C. SOWYER
First Vice-President   Treasurer

Great Kanawha Valley Improvement Association[97]
CHARLESTON-ON-KANAWHA, WEST VIRGINIA

October 13th, 1 9 2 8

The Public Service Comm. Of W. Va.,
Charleston, West Virginia.

Gentlemen: Attention—Mr. A.B. Moore, Secty.

I have the engineers' report regarding authority to construct dms, ets. [sic], of the New Kanawha Power Company for which I thank you.

I have gone over this report with the utmost care and I think that I can say our Association concurs very heartily with Mr. Krebs in this report and will trust that this permit will be favorably acted upon by the commission at its earliest discretion.

         Very truly yours,

         GREAT KANAWHA VALLEY IMPROVEMENT ASSN.

BY: Ernest M. Merrill, President

EMM'b

The Campbell's Creek Coal Co.[98]
THE HATFIELD-CAMPBELL CREEK COAL CO.
THE HATFIELD-RELIANCE COAL CO.

---

97. Ibid.
98. Ibid.

General Office-Union Trust Building
CINCINNATI

October 23, 1928

Mr. A.B. Moore, Secretary,
The Public Service Commission,
Charleston, West Virginia.

Dear Mr. Moore:

I beg to acknowledge receipt of your valued letter of October 18, advising that the Public Service Commission has fixed the date of October 26, at two o-clock P. M. for a hearing with reference to granting authority to the New-Kanawha Power Company to construct two dams.

My recollection is that these are the two dams that were under discussion in early September of this year, and I was there to enter protest in the event these dams would have any effect on the movement of water out of the New river[sic] that might effect navigation in the Kanawha river, and at the hearing I was satisfied that was not the case.

Therefore, unless there are some new developments or that the power dams under consideration are not the ones I heard discussed, if either these dams or any other dams have any effect on the flow of water into the Kanawha river [sic], then we should feel it our duty to protect navigation by entering our appearance and making protest; otherwise, we do not want to go to the expense and trouble of coming to the hearing. I might say I spent most of last week in West Virginia.

Yours very truly,

J.T. Hatfield.
CHAIRMAN OF THE BOARD.

JTH MP

MEMORANDUM OF
THE PUBLIC SERVICE COMMISSION OF WEST VIRGINIA
TO
THE FEDERAL POWER COMMISSION[99]

In the Matter of the Declaration Of New-Kanawha Power Company with Respect to the Proposed Development of Hydro-Electric Power on the New River Between Hawks Nest and Gauley Junction, In West Virginia.

---

99. Ibid.

The Public Service Commission of West Virginia having heretofore furnished the Federal Power Commission with copies of the application and general plans and drawings of New-Kanawha Power Company, a corporation, in Case no. 1863 . . . to construct, maintain and operate a dam, hydro-electric plant and tunnel on New River . . . known as Gauley Junction Dam, and . . . in Case No. 1864 an application for a permit for a similar hydro-electric development known as Hawks Nest Dam . . . .

. . . the Public Service Commission of West Virginia respectfully represents:

. . .that said Public Service Commission has assumed jurisdiction of the subject matter involved in the development of said hydro-electric works, and, pursuant to the duty laid upon it by the laws of said State, has fully investigated, among other things, the following subject in relation thereto:

(1) Whether or not the said New River is a navigable stream. . . .
(2) Whether or not the construction, maintenance and operation of said dams and works will affect the navigability of any other stream used and useful in intrastate, interstate or foreign commerce.

. . .such investigation resulted substantially in the following findings of fact:

(1) That said New River is not a navigable stream.
(2) That the construction, maintenance and operation of said dams and works . . . under such permit . . .will not result in any impairment of navigation upon the Great Kanawha River . . . .

. . .the Public Service Commission has issued permits . . . for the development of hydraulic power and hydro-electric energy for sale to the public for the period of fifty (50) years . . . [and] the New-Kanawha Power Company has voluntarily agreed and stipulated, among other things, that the State of West Virginia by its proper authority shall at all times have and freely exercise the powers, authority and jurisdiction to regulate and control the construction, equipment, maintenance and operation of said dam, hydro-electric plant and tunnel so as to conserve and protect all public and private rights in the waters of the state . . . .

Wherefore, the Public Service Commission of West Virginia respectfully urges . . . that the Federal Power Comission do not undertake to assume jurisdiction of said proposed hydro-electric development on said New River.

Respectfully submitted,

THE PUBLIC SERVICE COMMISSION OF WEST VIRGINIA
I. Wade Coffman, Chairman.

F.M. Livezey, Attorney.
Charleston, West Virginia, January 3, 1929

REQUEST TO MODIFY PLANS AND BUILD[100]
ONLY ONE DAM
AND ONE POWER PLANT

CASES NOS. 1864 AND 1863.

TO THE PUBLIC SERVICE COMMISSION OF WEST VIRGINIA:

Pursuant to the orders of the Commission entered in said cases on the 8th day of December, 1928, and the provisions of said orders retaining these cases on the docket and continuing the same for such future proceedings as are proper;

And further pursuant to the provisions of law and of said orders as to power and duty of the Commission to examine and pass upon detailed plans for the construction, equipment, maintenance, and operation of said dams and improvements, and such changes therein as shall be required or approved by the Commission;

The applicant, New-Kanawha Power Company, by L.H. Davis, its Vice-President, Owen M. Jones, its Chief Engineer, and Clyde B. Johnson, of its counsel, hereby respectfully presents to the Commission herein certain detailed drawings and plans of construction proposing certain modifications and changes with respect to the said two dams and projects, and the operation thereof, which changes and modifications it desires to make in the interest of economy and efficiency, and now moves the Commission to make an order approving and authorizing such changes.

The Campbell's Creek Coal Co.[101]
THE HATFIELD-CAMPBELL CREEK COAL CO.
THE HATFIELD-RELIANCE COAL CO.

General Office-Union Trust Building
CINCINNATI

September 18, 1929

Mr. Wade Coffman, Chairman,
The Public Service Commission of West Virginia,
Charleston, West Virginia.

---

100. Ibid.
101. Ibid.

Dear Mr. Coffman:

In this morning's mail, I have a letter from Mr. R.B. Bernheim, Secretary of the Commission, enclosing order for a hearing in the case of the New-Kanawha Power Company's application to construct a dam at Hawks Nest, the New-Kanawha Power Company making application for a change of location of the Plant and for modification in the location and general plans and drawings of The tunnel connecting said dam and hydro-electric plant, for the construction and operation of which dam, hydro-electric plant and tunnel a permit was granted in this case on the 8th day of December, 1928, also for authority to acquire certain additional lands and necessary rights in said additional lands described in said application and motion.

Mr. Coffman, in the first hearing for this permit, we attended it and, at a considerable expense, brought with us a number of experienced river men, namely, Captain Frank Johnson, who had been in the river business on the Great Kanawha river all his life, Captain Harry Martin, his father an old river man on the Great Kanawha river and he, himself, a Kanawha river captain and pilot of more than forty years experience. We brought Mr. W.V. Rensford from Dana, who is now Superintendent and who had been employed by the Campbell's Creek Coal Company for more than forty-eight years, Mr. W.C. Mitchell of Plymouth, West Virginia, and James Hatfield and myself.

As I view this application for the changes that are suggested, it is a complete reversal for new plans, new location and new whatnot. Our experience is this, without intending to attribute to the New-Kanawha Power Company or to Mr. Clyde B. Johnson, its counsel, any modifications or wishes or desires not in conformity with this first application, I do say that it is not unusual for corporations and counsel to proceed in this matter to confuse and mix up the issues to the extent that they get what they want at the expense of the coal companies and transportation companies on the Great Kanawha River. Please get me straight – it would be foolish for me to say that this is a fact, but it is sensible to say that it is the usual procedure to mix, confuse and twist.

We wish to go on record with this statement that a single inch of water denied the coal companies and transportation companies on the Great Kanawha river [sic] by the erection of a power plant, to even that small change in the amount of water, we most earnestly protest against, and for the following reasons:

1. The New and Great Kanawha rivers were God-given to the people of West Virginia.

2. The coal interests and the transportation interests are far more to the State of West Virginia than any hydro-electric plant owned and controlled by foreign capital and its merchandise distributed, to a large extent, in foreign states.

3. No plant should receive a commission from the State of West Virginia for the producing of electrical current that has the possibility in the way of operation or control by that company of having any effect on the usual flow of water of the New and Kanawha rivers.

Therefore, at this time, we offer a protest in the name of our company, part of which has been in existence for more than sixty years (the old Campbell's Creek Coal Company) and the Hatfield interests for more than forty-seven years, which operates three mines on Campbell's Creek producing more than 3500 tons a day and a mine at Plymouth, West Virginia, producing more than 1000 tons a day, and transports and moves more coal and more products than any other company on the Kanawha river, and, having been in business all these years, we now protest against any changes of location, any new methods of operation, any storages or restricted flow of the river than can have an affect on the river to as much as one inch, because one inch of water denied the transportation interests in many cases prevents them from moving their products.

I do not know that I will attend your meeting. It seems to me that all the time and expense we went to in bringing experts there, that it should not be necessary for us to do it all over again, but this letter will be our protest against any changes in the former plans as presented.

I am sending a copy of this letter to Governor Conley, and this for your information, for the reason that I believe him to understand the coal and transportation interests very fully, and I wish him to know our position in the matter.

I wish to express to you, Mr. Coffman, and to the other members of the Committee my remembrances of the courteous treatment accorded me when I appeared before the Commission. With very kindest regards to yourself, and begging that you will pass to the other members of the Committee and the Secretary my kind remembrances, I am,

>Yours very truly,
>J.T. Hatfield
>CHAIRMAN OF THE BOARD.

JTH MP
CC—Gov. Wm. G. Conley, Charleston, W. Va.

According to records archived in the PSC repository in Charleston, WV, **J.E. Settle** served at the will of the WV Public Service Commission in overseeing construction of the Hawks Nest Project. The following observations have been excerpted from a few of his numerous reports. The PSC relieved Settle of his employment on July 31, 1933. He was succeeded by Charles A. Ray.[102]

---

102. Ibid.

12/15/1930.

Gentlemen:-

I have delayed making a report on construction work on account of necessary additional excavation at the dam site, which I will describe later in detail.

The excavation in the tunnel is preceeding [sic] satisfactorily at all four headings, the total amount of tunnel driven to date being 4500 feet.

In previous reports I have called the Commission's attention to the fact that the rock excavated from the lower or outlet end of tunnel is all being crushed and used for concrete at the dam and being transported to Boncar. The rock being taken to Boncar will be used in one of their factories as it contains a very high percentage of silica and from information this will be used in the manufacture of Ferro-silicon. My reason for calling this to your attention is that the tunnel at this point has been enlarged to 42 feet diameter to procure the additional material for the use of the plant at Boncar. [remainder omitted]

1/7/1931

Core drilling and sampling of the sandstone through which the tunnel connecting the power house at Gauley Junction and New River and the diversion dam at Hawks Nest on New River is driven for a great part of the distance of 16,000 feet showed a high silica sandstone, and tests of this material crushed showed it to be acceptable for concrete aggregates. These tests were made by both the Company and outside testing laboratories. [remainder of main report omitted]

July, 1931
SUPPLEMENTAL REPORT

As a part of the record of this project mention must be made of certain investigations into a prevalence of pneumonia among the employees of Rhinehart [sic] and Dennis.

The housing conditions are briefly described as three construction camps numbered One, Two and Three. Camp One houses majority of employees of power house and No. 1 heading. Camp Two houses all the workmen at the dam and No. 4 heading. Also there are numerous men living around in the vicinity of Hawks Nest, Cotton Hill, Ansted, Gauley Bridge and nearby communities who are not living in company houses. These are ordinary type construction houses of average good conditions for temporary living quarters. Physicians are living in camps, one at Camp Three and one at Glen Ferris, in constant touch with the work. All cases requiring attention are cared for at Coal Valley Hospital, Montgomery, W. Va., when found in any of these camps.

Approximately 80% of more of all employees of Rhinehart and Dennis at present employed, or formerly employed, are colored. A large percent is the floating construction type of negro from the Carolinas, Tennessee and Virginia. Labor turnover is heavy. No medical examination for fitness for physical labor is used and at a time such as the present period of depression with work scarce numbers of these men reach their destination on feet and weakened from exposure and poor nourishment. A check is being made of the labor turnover at present time but I estimate a probable maximum of 4,000 men have been employed within the part year or thirteen months of construction. Living habits and conditions, poor resistance and change of climate should be mentioned as contributory causes to the prevalence of pneumonia in the camps.

Attention of writer was called to this condition during the latter part of the winter, several deaths having resulted from pneumonia. The weather was very disagreeable, and being familiar with the low resistance of Negroes to various diseases of this kind, the writer did not deem it within his jurisdiction to report same. Local rumors and newspaper publicity of the extent of the prevalence of pneumonia with existing fatalities and other fatalities resulted in Mr. Lambie, Chief of Department of Mines, being called by local authorities for the investigation of conditions in tunnel. His cooperation and suggestions were appreciated. An additional fan suggested for #1 heading was placed at the adit for exhausting the air at the feet [sic] of the slope, and the suggestions mentioned heretofore were made to contractor who complied immediately with his request.

The information gathered by the writer shows quite a heavy record of pneumonia. Approximately one hundred cases were sent in from camps and vicinity over a period of 13 months. 27 of these were fatal. Ninety percent were negroes.

On this account the writer spent several days on this work investigating conditions, accompanied on one trip by Mr. Lambie personally, and at other times by officers of contractor, engineers and inspectors of New Kanawha Power Co., or alone.

The writer has copies of reports of State Department of Mines, New Kanawha Power Co., the Montgomery Clinic and has interviewed and carefully questioned the local physicians. Analyses have been made of the air in the tunnel, tests have been made of the blood of employees subject to the conditions prevailing.

The air samples show quite a difference in carbon monoxide content. Some of the worst samples show dangerous concentrations if breathed for several hours, others practically none. Sampling has been done and analyses made by different persons at different times. No direct instances of carbon monoxide poisoning have been observed by writer on any of his many trips into the different headings (such as would result in immediate sickness or nausea). I have carefully watched for this since the date of Mr. Lambie's report on the condition of air in tunnels by his representative, Mr. Sims. No ill effects have ever been experienced by the writer on visits exceeding two hours in each tunnel or as high as six hours daily. There is almost a complete absence of dust, due to wet drilling, but numbers of these men are reported to go home in wet clothes and sleep in them. Study of carbon monoxide literature U.S. Health Bulletin 195 (A Review of Carbon Monoxide Poisoning) mentions some authority where dust and carbon monoxide are contributory causes of pneumonia. Others do not agree.

[omission]

The relation of possible carbon monoxide concentration to pneumonia is not within the engineering jurisdiction contemplated in this supervision but for purposes of record am giving you herewith the report of the Montgomery Clinic, as an opinion of the medical profession familiar with the conditions and treatment of these cases under their supervision and observation.

Respectfully submitted,

J.E. Settle
Engineer for the Commission

[The following letter from W. V. Wilkerson, M. D., attending physician at the Coal Valley Hospital in Montgomery, was included in Settle's report of July, 1931.]

MONTGOMERY CLINIC
Montgomery
W. Va.

July 3, 1931.

Dr. William R. Laird,
Montgomery, W. Va.

Dear Dr. Laird:

In regard to your inquiry relative to the reason for the epidemic of pneumonia among the employees of the Rinehart & Dennis Company, In., near Gauley Bridge, W. Va., I feel that perhaps a combination of factors may be responsible.

It is well known that contact infection is the usual cause of spread of pneumonia and I believe this to be the major factor here. This belief is based upon the following facts.

(1) Early in this epidemic there were a great many Type II infections coming only from Camp No. 2 with Type IV infections coming occasionally from Camp No. 3. After a period of time and type II cases ceased and we began to receive Type I and Type IV infections from the various camps, it appearing evident that the specific organism was being transmitted from person to person.

(2) We have had quite a few cases of pneumonia from "hanger-on" at the camps who have not been employed and who, therefore, have not been exposed to any factors in the tunnel that may be contributing to the pneumonia rate. At the present time we have a colored woman in the Hospital with pneumonia.

A high concentration of carbon monoxide has been reported by Cecil and others to be responsible for an increased pneumonia rate. Our examination of the blood of these patients to determine the percentage of carbon monoxide present has given results that are so nearly negative that I am not inclined to attach any particular significance to this factor.

I have not been able to see any evidence in the literature to sustain the opinion that silicosis increases the pneumonia rate.

I feel that the rather poor living condition of these patients may contribute to the incidence of pneumonia through lowered resistance.

A further factor to be considered is the fact that some of these patients do not quit work as soon as they become ill but continue to work during the early stages of pneumonia, thus serving as an active source from which the infection may spread. I now have in the Hospital one man who worked for four days from the onset of pain in the chest, chill, and blood-tinged sputum.

Summarizing the above, I believe that contact is the most important factor responsible for this pneumonia epidemic. Crowded living conditions undoubtedly play an important part in these contact infections not only by direct infection but by causing a high percentage of "carriers." The fact that these patients frequently continue to work after the onset of illness is undoubtedly an important factor. I regard the amount of carbon monoxide these patients show and the presence of silicosis as minor factors only.

Respectfully submitted,
W.V. Wilkerson, M.D.

Dr. Wilkerson considered the living quarters of the laborers far more conducive to the "pneumonia epidemic" than the work conditions in the tunnel. In 1931 he diagnosed "pneumonia" as the "cause of death" for twenty-four out of the twenty-six deaths he attended at the Coal Valley Hospital. Of the remaining two, one was determined to be "tuberculosis" and the other "hypertonsial & arterioscleric cardio renal disease—starvation, acute malnutrition." Time and again comments on death certificates signed by Dr. Wilkerson included: "Living in construction camp—very unsatisfactory living conditions," and even on one occasion, answering the question on the death certificate "was the occupation of the deceased in any way related to his death?" Dr. Wilkerson wrote, "Yes—crowded living quarters."

3/20/1931

[omission]

The total heading driven to date is 8500 feet or slightly in excess of one-half the total length of the project. Heading No. One, or Power House end, has advanced 2700 feet and character of spoil has not changed, still being crushed for dam construction and for Boncar plant, and is satisfactory. Diameter 43 feet."

[omission]

In respect to employment and wager there are an average of 1,000 men employed on this project. The minimum wage for common labor is .30 per hour. Heading men and drillers .35. Carpenters and form setters .50 per hour. Two shifts of 9 to 11 hours are worked in tunnel headings."

7/23/1931

[omission]

"To date this project has had 9 fatal accidents. Two of these were engineers sounding the river before the contractor, Rhinehart [sic] & Dennis, came on the work. The seven fatal accidents were 5 inside the tunnel and 2 outside. Of the 2 outside one man was killed in shafting of crusher at power house. One was drowned at dam site. The five fatalities in tunnel were one caught by steam shovel against rib, one fell off trip cars (reported riding against rules0, and the other three were due to rock or roof falls. The roof fall record is exceptionally good considering the hazard of working from 15 to 35 feet underneath top where the fall of a small sized fragment of stone might be fatal. Steel timbering and lagging were abandoned in #4 heading at 4500 feet and all headings are now in Sandstone. The grade of sandstone found in #3 and #4 headings is not as good as in #1 and #2 for use at Boncar, so this spoil is wasted.

An average of 60 men are working in each heading or 480 men each 24 hours. Counting 330 working days since start of work, we have approximately 1,584,000 man hours. With a labor turnover of 3 to 1 or approximately 1500 men used on tunnel work the fatalities from roof falls are one to each 500,000 man hours, or one in 500 men employed.

[omission]

The ventilation at #1 [Power House] heading is by a fan located back of blacksmith shop, this being the only practical place to install it on account of shooting in the penstock of power house. The contractor at the suggestion of Mr. Lambie, Chief of the State Department of Mines, who has cooperated with the writer in checking ventilating conditions at the various tunnel headings, and myself, has placed the switch for this blower fan on the outside of shop accessible from both inside and outside, day or night. . . .

[omission]

Haulage at #1 heading is now being handled by one gasoline locomotive, one being removed at our suggestion. As the haulage can be handled by one locomotive (probably not as efficiently) but it removes from the tunnel one half the danger from carbon monoxide from the exhaust of the locomotive.

[omission]

11/30/1932

...and while no definite program f amount of tunnel to be lined has been agreed on the writer expects all the small bore section to be lined complete, and also the floor of the large section. The power house structure (outside walls) is completed and is now entirely enclosed. The inside finishing and setting of panels and wiring and all inside finishing can now be done under cover from the winter weather.

The excavation for the tailrace is practically completed.

[omission]

While I have given personal notice to the commission that due to the suits instituted against Rhinehardt [sic] and Dennis the contractors on the project, all estimates are now held up, and all equipment has been attached pending the outcome of the damage suits on account of silicosis pending in Fayette County, think it should be noted here as a part of the record on this project."

[omission]

On July 8, 1933 there were 225 men employed on the entire project. Common labor was being paid twenty cents per hour for outside work. Carpenters and steel men 45 cents with higher prices for skilled labor.

[omission]

Settlement is now being made of the damage suits filed against Rhinehardt [sic] and Dennis from alleged contraction of silicosis in the conduct of their tunnel operations. This will be reported in detail after all releases are signed. All work on this project has conformed to existing plans and specifications and no changes have been made in the personnel of the contractors or of the engineering and inspection forces of the New Kanawha Power Co.

LAW OFFICES[103]
CLYDE BEECHER JOHNSON
CHARLESTON, WEST. VA.

August 5, 1933

The Public Service Commission
Charleston, West Virginia

---

103. Ibid.

Gentlemen: <u>Attention Mr. Jos. G. Conley, Secretary</u>

Thanks for copy of the Commission's letter of August 4, addressed to Chief Engineer O.M. Jones of New-Kanawha Power Copmpany, advising that Mr. J.E. Settle, engineer and inspector of the dam and works formerly owned by New-Kanawha Power Company at Hawks Nest, and whose employment in this behalf ended with July 31, 1933, has been succeeded by Mr. Charles A. Ray, who has been assigned to represent the Commission as its engineer and inspector of the Hawks Nest project.

I am sure that Mr. Jones will know Mr. Ray and be able to advise as to his qualifications. I am certain that the officials of Electro Metallurgical Company, who will now carry the burden of this cost will be pleased at the action of the Commission in reducing the cost of this engineer supervision. However, it was my recollection that New-Kanawha Power Company has been assessed about $5,000.00 per annum to cover the cost of this inspection, but since discussing this matter over the telephone with Judge Nethken, I realize that I am probably in error as to this, and that the $5,000.00 was not for each year for such time as the assessment might last and that a new assessment would be made only when the former assessment was exhausted.

I am sending copy of your letter of the 4th inst. To Mr. Benjamin G. Smith, who has been counsel for the New-Kanawha and will, I suppose, act in the same capacity for the Electro Metallurgical Company. You will doubtless hear from Mr. Jones and Mr. Smith in reply to your letter.

Very truly yours,
Clyde B. Johnson

U.S. Supreme Court
UNITED STATES v. STATE OF WEST VIRGINIA, 295 U.S. 463 (1935)

295 U.S. 463

UNITED STATES

v.

STATE OF WEST VIRGINIA et al.

No. 17.

Argued May 2, 1935.
Decided May 20, 1935.

[295 U.S. 463, 465]   Mr. Homer A. Holt, of Charleston, W. Va., for defendant State of West Virginia.

Messrs. Edward W. Knight and Robert S. Spillman, both of Charleston, W. Va., for defendants Electro Metallurgical Co. and others.

[295 U.S. 463, 467]

Mr. Justice STONE delivered the opinion of the Court.

This is an original suit in equity, brought by the United States, in which relief by injunction is sought against the defendants, the state of West Virginia, Union Carbide & Carbon Corporation, a New York corporation, and its wholly owned subsidiaries, Electro Metallurgical Company and New-Kanawha Power Company, West Virginia corporations. The questions now presented are raised by separate motions, one by the state of West Virginia, the other by the corporate defendants, to dismiss the bill of complaint on the grounds that it does not state any justicable controversy between the United States and the state of West Virginia, and that it appears upon the face of the bill of complaint that this Court has no original jurisdiction of the suit against the defendants or any of them.

The bill of complaint, filed January 14, 1935, contains allegations which so far as now relevant, may be detailed as follows: The New river flows northwesterly across the state of West Virginia and near the center of the state joins the Gauley river to form the Kanawha river, which flows thence to the state boundary and into the Ohio river. The New and Kanawha rivers are one continuous interstate stream, which throughout its course constitutes navigable waters of the United States.

There are many locations for dam sites on the rivers; four dams have been constructed on the New river at points in Virginia and West Virginia, and a fifth at Hawks Nest, W. Va., upon which the present litigation centers, [295 U.S. 463, 468] is now approaching completion.

The United States has constructed ten dams on the Kanawha river for the purpose of improving navigation, and is now engaged in construction work on two additional dams on the Kanawha river immediately below the Hawks Nest project, and has in contemplation the construction of a large reservoir at Bluestone, W. Va., on the New river above the Hawks Nest project, for purposes of flood control, production of power, and in aid of navigation.

It is alleged that the New and Kanawha rivers throughout West Virginia constitute a continuous stream which was in its natural condition and still is susceptible of navigation, and is a highway capable of being improved and used for purposes of interstate and foreign commerce; that any obstructions to its navigability will be removed or overcome by improvements initiated by the United States and now in operation or in the course of construction; that the Hawks Nest project will seriously obstruct navigation in the New and Kanawha rivers, by producing fluctuations in the flow of New river; and that, upon the filing by New-Kanawha Power Company of a declaration of intention to construct the dam, pursuant to section 23 of the Federal Water Power Act, chapter 285, 41 Stat. 1063, 1075, 23, 30, 16 U.S.C. 791, 817

(16 USCA 791, 817), <u>the Federal Power Commission determined that the proposed Hawks Nest dam would affect the interests of interstate commerce, and that under the act the dam could not lawfully be built without a license from the Commission.</u>

It is further alleged that the defendant New-Kanawha Power Company has obtained from the Public Service Commission of West Virginia a license or permit to construct the dam at Hawks Nest for power purposes. This permit was later transferred to the defendant Electro Metallurgical Company; and the corporate defendants, acting under the state license, are now engaged in the construction of the dam. It is alleged that its construction [295 U.S. 463, 469] is in violation of the Act of Congress of March 3, 1899, c. 425, 9, 30 Stat. 1121, 1151, 33 U.S.C. 401 (33 USCA 401), and the Federal Water Power Act, in that the plans for the project have not received the consent of Congress or the approval of the Chief of Engineers of the United States Army and the Secretary of War, and the defendants have received no license for the project from the Federal Power Commission.

The allegations with respect to the state of West Virginia are that the state challenges and denies the claim of the United States that the New river is a navigable stream; that the state asserts a right superior to that of the United States to license the use of the New and Kanawha rivers for the production and sale of hydroelectric power, and denies the right of the Federal Power Commission to require a license for the construction and operation of the Hawks Nest project by the corporate defendants; and that the state asserts that, in so far as the Federal Water Power Act purports to confer upon the Federal Power Commission authority to license the project or to control the use of the river by the corporate defendants, the act is an invasion of the sovereign rights of the state and a violation of the Constitution of the United States. The bill further elaborates, in great detail and particularity, but does not enlarge, these basic allegations.

It prays an injunction restraining the corporate defendants from constructing or operating the Hawks Nest project without a license from the Federal Power Commission. It also asks an adjudication that the New river is navigable waters of the United States and that the United States has the right to construct and operate, and to license others to construct and operate, dams and connected hydroelectric plants on the New and Kanawha rivers. We are asked to declare that any right of the state of West Virginia to license the construction and operation of dams upon the rivers, or to sell or to license [295 U.S. 463, 470] others to sell power generated at such dams, is subject to the rights of the United States, and to enjoin the state from asserting any right, title, or interest in any dam, or hydroelectric plant in connection with it, or in the production and sale of hydroelectric power on the New and Kanawha rivers, superior or adverse to that of the United States, and from in any manner disturbing or interfering with the possession, use, and enjoyment of such right by the United States.

It can no longer be doubted that the original jurisdiction given to this Court by section 2, art. 3 of the Constitution, in cases 'in which a state shall be a party,' includes cases brought by the United States against a state. United States v. Texas, 143 U.S. 621,

12 S.Ct. 488; United States v. Michigan, 190 U.S. 379, 396, 23 S.Ct. 742; Oklahoma v. Texas, 252 U.S. 372, 40 S.Ct. 353; Id., 258 U.S. 574, 581, 42 S.Ct. 406; United States v. State of Minnesota, 270 U.S. 181, 195, 46 S.Ct. 298; United States v. State of Utah, 283 U.S. 64 , 51 S.Ct. 438; compare Florida v. Georgia, 17 How. 478, 494; United States v. North Carolina, 136 U.S. 211, 10 S.Ct. 920.

But the original jurisdiction thus conferred is only of those cases within the judicial power of the United States which, under the first clause of section 2, art. 3 of the Constitution, extends 'to all Cases, in Law and Equity, arising under this Constitution, the Laws of the United States, and ... to Controversies to which the United States shall be a Party ...' Com. of Massachusetts v. Mellon, 262 U.S. 447, 480-485, 43 S.Ct. 597; see State of Wisconsin v. Pelican Insurance Co., 127 U.S. 265, 289, 8 S.Ct. 1370. Our original jurisdiction does not include suits of the United States against persons or corporations alone (see In re Barry, 2 How. 65; Louisiana v. Texas, 176 U.S. 1, 16, 20 S.Ct. 251; Baltimore & O.R. Co. v. Interstate Commerce Commission, 215 U.S. 216, 224, 30 S.Ct. 86; State of Oklahoma v. Texas, 258 U.S. 574, 581, 42 S.Ct. 406), nor is it enough to sustain the jurisdiction in such a case that a state has been made a party defendant. The bill of complaint must also present a [295 U.S. 463, 471] 'case' or 'controversy' to which the state is a party, and which is within the judicial power granted by the judiciary article of the Constitution.

Hence we pass directly to the question whether the bill of complaint presents a case or controversy between the United States and the state of West Virginia within the judicial power. The answer is unaffected by the fact, set forth in the bill of complaint, that the state, on its application to intervene in a suit, since discontinued, brought by the United States, in the District Court for West Virginia to restrain the corporate defendants from constructing the dam, asserted its interest as a state in the development of power under state license at the Hawks Nest dam, particularly in the license fees and taxes to be derived from the project. The details of the attempted intervention at most serve only to support the allegations of the bill, that the state has asserted the right, through a license of the Hawks Nest project, to control the use of the rivers for power purposes.

At the outset, it should be noted that the bill in the present suit neither asks the protection nor alleges the invasion of any property right. It asserts no title in the United States to the bed of the stream, which might afford a basis for a suit to remove a could on title, as in United States v. State of Utah, supra, and United States v. State of Oregon, 295 U.S. 1, 55 S.Ct. 610, 79 L.Ed. —, decided April 1, 1935. It alleges that the United States has built dams on the Kanawha river below the Hawks Nest project, and has acquired lands in pursuance of its plans for flood control, improvement of navigation, and the generation and sale of hydroelectric power on both rivers. But there is no allegation of any interference by the state, actual or threatened, with any of the land or property thus acquired.

The only right or interest asserted in behalf of the United States is its authority under the Constitution to [295 U.S. 463, 472] control navigable waters, and particularly the right to exercise that authority through the Federal Power Commission. Since that authority is predicated upon the single fact, fully alleged in the bill and admitted by the motions to dismiss, that the rivers are navigable waters of the United States, the power of the United States to control navigation, and to prevent interference with it, by the construction of a dam except in conformity to the statutes of the United States, must be taken to be conceded. See State of New Jersey v. Sargent, 269 U.S. 328, 337, 46 S.Ct. 122.

But the bill alleges no act or threat of interference by the state with the navigable capacity of the rivers, or with the exercise of the authority claimed by the United States or in behalf of the Federal Power Commission. It alleges only that the state has assented to the construction of the dam by its formal permit, under which the corporate defendants are acting. There is no allegation that the state is participating or aiding in any way in the construction of the dam or in any interference with navigation; or that it is exercising any control over the corporate defendants in the construction of the dam; or that it has directed the construction of the dam in an unlawful manner, or without a license from the Federal Power Commission; or has issued any permit which is incompatible with the Federal Water Power Act; or, indeed, that the state proposes to grant other licenses, or to take any other action in the future.

Section 28 of the Water Power Act of West Virginia, chapter 17 of the Acts of 1915, which gives to the state Public Service Commission its authority, provides that: 'Nothing contained in this act shall be construed to interfere with the exercise of jurisdiction by the government of the United States over navigable streams.' The bill seeks an injunction, against the corporate defendants, restraining only the construction of the dam without a [295 U.S. 463, 473] license from the Federal Power Commission.

But section 9(b) of the Federal Water Power Act (16 USCA 802(b) requires that every applicant for a license shall present 'satisfactory evidence that the applicant has complied with the requirements of the laws of the State or States within which the proposed project is to be located with respect to bed and banks and to the appropriation, diversion, and use of water for power purposes and with respect to the right to engage in the business of developing, transmitting, and distributing power. ...' The mere grant of the state license, which the Federal Water Power Act makes prerequisite to the application for the federal license, cannot be said to involve any infringement of the federal authority. It does not appear that the state has done more.

We any assume, for present purposes, that the United States as sovereign has a sufficient interest in the maintenance of its control over navigable waters, and in the enforcement of the Federal Water Power Act, to enable it to maintain a suit in equity to restrain threatened unlawful invasions of its authority (see State of Kansas v. Colorado, 185 U.S. 125, 22 S.Ct. 552; State of Georgia v. Tennessee Copper Co., 206 U.S. 230, 237, 27 S.Ct. 618, 11 Ann.Cas. 488; Marshall Dental Manufacturing Co. v. State of Iowa, 226 U.S. 460, 462, 33 S.Ct. 168; State of Missouri v. Holland, 252 U.S. 416,

431, 40 S.Ct. 382, 11 A.L.R. 984; see Hudson County Water Co. v. McCarter, 209 U.S. 349, 355, 28 S.Ct. 529, 14 Ann. Cas. 560), and that a cause of action within the jurisdiction of a federal District Court is stated against the corporate defendants who are alleged to be engaged in building an obstruction in navigable waters of the United States.

But there is presented here, as respects the state, no case of an actual or threatened interference with the authority of the United States. At most, the bill states a difference of opinion between the officials of the two governments, whether the rivers are navigable and, consequently, whether there is power and authority in the [295 U.S. 463, 474] federal government to control their navigation, and particularly to prevent or control the construction of the Hawks Nest dam, and hence whether a license of the Federal Power Commission is prerequisite to its construction. There is no support for the contention that the judicial power extends to the adjudication of such differences of opinion. Only when they become the subject of controversy in the constitutional sense are they susceptible of judicial determination.

See Nashville, Chattanooga & St. Louis R. Co. v. Wallace, 288 U.S. 249, 259, 53 S.Ct. 345, 87 A.L.R. 1191. Until the right asserted is threatened with invasion by acts of the state, which serve both to define the controversy and to establish its existence in the judicial sense, there is no question presented which is justiciable by a federal court. See Fairchild v. Hughes, 258 U.S. 126, 129 , 130 S., 42 S.Ct. 274; State ex rel. Texas v. Interstate Commerce Commission, 258 U.S. 158, 162, 42 S.Ct. 261; Com. of Massachusetts v. Mellon, supra, 262 U.S. 447, 483-485, 43 S. Ct. 597; State of New Jersey v. Sargent, supra, 269 U.S. 328, 339, 340 S., 46 S.Ct. 122.

General allegations that the state challenges the claim of the United States that the rivers are navigable, and asserts a right superior to that of the United States to license their use for power production, raise an issue too vague and ill-defined to admit of judicial determination. They afford no basis for an injunction perpetually restraining the state from asserting any interest superior or adverse to that of the United States in any dam on the rivers, or in hydroelectric plants in connection with them, or in the production and sale of hydroelectric power. The bill fails to disclose any existing controversy within the range of judicial power. See State of New Jersey v. Sargent, supra, 269 U.S. 328, 339, 340 S., 46 S.Ct. 122.

The government places its chief reliance upon the decision in United States v. State of Utah, supra, in which this Court took original jurisdiction of a suit, brought by the United States against the state, to quiet title to the bed of the [295 U.S. 463, 475] Colorado river. But the issue presented by adverse claims of title to identified land is a case or controversy traditionally within the jurisdiction of courts of equity. Such an issue does not want in definition. The public assertion of the adverse claim by a defendant out of possession is itself an invasion of the property interest asserted by the plaintiff, against which equity alone can afford protection. See United States v. Oregon, supra. A different issue, in point of definition of threatened injury and imminence of the controversy, is presented by rival claims of sovereign power made by the national and a state government.

The sovereign rights of the United States to control navigation are not invaded or even threatened by mere assertions. It is, in this respect, in a position different from that of a property owner, who because of the adverse claims to ownership can neither sell his property nor be assured of continued possession. The control of navigation by the United States may be threatened by the imminent construction of the dam, but not by permission to construct it.

No effort is made by the government to sustain the bill under the Declaratory Judgment Act of June 14, 1934, c. 512, 48 Stat. 955 (28 USCA 400). It is enough that that act is applicable only 'in cases of actual controversy.' It does not purport to alter the character of the controversies which are the subject of the judicial power under the Constitution. See Nashville, Chattanooga & St. Louis R. Co. v. Wallace, supra.

Since there is no justiciable controversy between the United States and the state of West Virginia, the cause is not within the original jurisdiction of this Court, and must be dismissed.

It is so ordered.

Mr. Justice BRANDEIS is of opinion that the United States should be granted leave to amend its bill.

TAX DEPARTMENT[104]
30 East 42nd Street
New York

June 8, 1932.

Mr. E.V. Williamson, Statistician,
The Public Service Commission of West Virginia,
Charleston, W. Va.

Dear Sir:

In reply to your letter of May 31st, 1932, with respect to the change reflected in the Annual Report for 1931 of New-Kanawha Power Company, permit me to advise that it is my understanding that the change in the financial structure was occasioned because of the necessity of complying with a legal technicality.

The site of the hydro development had not been transferred to New-Kanawha

---

104. Ibid.

Power Company and consequently improvements made by the Power Company became, because of their attachment to the land, the property of the owner of the land. To correct this situation the Power Company was reimbursed by the owner of the land for the expenditures the Power Company had made for such improvements.

Although, as you understand, the transactions are inter-company, it was thought best to observe the technical legal requirements of the situation.

>   Very truly yours,
>   P. E. Katz

THE PUBLIC SERVICE COMMISSION[105]
OF WEST VIRGINIA
CHARLESTON

June 13, 1932.

CASE NO. 1864
NEW-KANAWHA POWER COMPANY,
Hawks Nest Hydro-Electric Project.

Dear Sir:

I think we have discussed with you the failure of New-Kanawha Power Company to ask approval of its stock issues as provided by section 33 of the Act under which its license was granted in the above proceeding, also its failure to submit the price and terms under which it proposed to purchase the lands designated in the permit (16th An. Rpt. 51, 80).

Recently it has come to our attention through the company's annual report for 1931 that it no longer carries the cost of the Hawks Nest Project in the Fixed Capital Account. Mr. Williamson inquired about this matter and received a letter under date of June 8, 1932, a copy of which is enclosed herewith.

In view of the whole situation the Commission feels the company should be required to conform to the provisions of the license in these respects promptly, and I am enclosing herewith draft of a letter, without date, which we have in mind forwarding to counsel for the power company, but we would like for you to go over it carefully and make any suggestions which may occur to you before the letter is dated and forwarded.

>   Yours very truly,
>   Chairman.

---

105. Ibid.

[To] Mr. F.M. Livesey,
Counsel, Public Service Commission,
Huntington, West Virginia.

THE PUBLIC SERVICE COMMISSION[106]
OF WEST VIRGINIA
CHARLESTON

June 17, 1932.

CASE NO. 1864
NEW-KANAWHA POWER COMPANY,
Hawks Nest Hydro-Electric Project,

Dear Sir:

I am authorized by the Commission to call your attention as counsel for New-Kanawha Power Company to what appears to be neglect on the part of that company to comply with certain provisions of the permit or license granted it December 8, 1928, in the above proceeding.

The permit authorizes the construction and operation of the project subject to the provisions of the then existing law (Water Power Act 1915), and includes approval of the acquisition of certain lands as provided by section 17 of the Act, but subject, if such acquisition be by purchase, to Commission approval of the price and terms. Among other provisions of that Act are, 1st, the provision in section 33 for the approval by the Commission of all issues of capital stock, bonds and other securities, and, 2nd, the provision of section 36 that the transfer, sale or assignment of any property of a licensee must be approved by the Commission.

It is recalled that the proviso in the license as to approval of the price and terms at which land might be acquired, and the requirements of section 33, were called to the attention of counsel, and that subsequently, in May, 1931, the company filed its election to accept the provisions of the 1929 Water Power Act. That election, of course, is void because of the invalidity of the 1929 Act.

The Commission has had it in mind for some time to call attention to the matter of the purchase of the lands designated in the permit, and that of approval of stock issues, and recently our Statistical Department has learned from the company's report for 1931 that the project in course of construction at Hawks Nest is no longer carried in the company's capital account. Upon inquiry by our Statistician, Mr. P.E. Katz, of the

---

106. Ibid.

company's Tax Department, advises that "the site of the hydro development had not been transferred to New-Kanawha Power Company and consequently improvements made by the Power Company became, because of their attachment to the land, the property of the owner of the land," and that the landowner has reimbursed the Power Company for the cost of the improvements. It appears, therefore, that an attempt has been made by the licensee to transfer these property rights without regard to the provisions of section 36 of the ACT.

In view of the confusion into which further neglect of the provisions of the license is leading, we urge the Company to give immediate attention to the matter of, 1st, the provision concerning the purchase of land, 2nd, the necessity for approval of stock issues, and 3rd, the revision of its accounts to show the cost of the project, the transfer of which to the landowner was attempted without the approval provided by section 36 of the Act.

>Yours very truly,
>Chairman.
>[to] Mr. Clyde B. Johnson,
>Charleston, West Virginia.

BENJAMIN GREGORY SMITH[107]
Attorney and Counselor at Law
30 East Forty-Second Street
New York

June 24, 1932.

Hon. Clyde B. Johnson,
Kanawha Valley Building,
Charleston, W. Va.

Dear Senator Johnson:

We are in receipt of your letter of June 17, 1932 and the letter of June 17, 1932 from Mr. Coffman to you, enclosed with your letter, with reference to the Hawks Nest project of New-Kanawha Power Company.

Since the decision of the Supreme Court of Appeals of West Virginia on August 13, 1931 holding absolutely void the West Virginia water Power Act of 1929, we have awaited a decision by the Public Service Commission as to the status of our permits after the said decision by the West Virginia court. As you know, we have always been

---

107. Ibid.

zealous to comply with the laws of West Virginia and in all respects to conduct our operations under the permits in exactly the manner desired by the Commission. It seemed to us appropriate that any decision involving the interpretation of the present status of our permits should originate with the State or through its legally constituted body for such purpose, The Public Service Commission.

From the said copy of the letter of June 17, 1932 from Mr. Coffman to you in which he states that he is authorized by the Commission to call your attention to certain matters, it would seem that the Commission has officially elected to hold that the West Virginia Water Power Act of 1915 is now in effect in so far as the permits of New-Kanawha Power Company are concerned. We are not disposed to question the decision of the Commission and we shall, of course, as we have always in the past, endeavor to comply fully with the wishes of the Commission.

Our latest hearing before the Commission, of which I have record, was a hearing held on June 2, 1931, at which hearing evidence was introduced to aid the Commission in computing the annual rate to be charged under the permit. At that time it was assumed by the Commission and by us that the 1929 act was valid, and we theretofore filed with the Commission an election that the permit for Hawks Nest should go under the 1929 act. At that time the Commission was of the opinion, because of such election so filed, that it was not necessary to introduce evidence of the financing of New-Kanawha Power Company. So far as my records show, we have not, since the hearing of June 2, 1931 until the said letter of June 17, 1932, received from the Commission any suggestion that a formal hearing be had.

You can assure the Commission that we feel that there has been no neglect on our part, and if we have not furnished to the Commission all the information that the Commission desires, it has been because we did not know that the Commission required such further information at any particular time.

I think that the questions raised by the Commission can be readily and satisfactorily answered.

From your letter and the letter of Mr. Coffman it would appear that the points raised by the Commission are these:

1. Approval by the Commission is required of all issues of capital stock, bonds and other securities.

2. Acquisition of lands is subject to the approval of the Commission.

3. Any sale of the property of the permittee is not valid unless approved by the Commission.

---

1. New-Kanawha Power Company has issued 40,000 shares of stock at $25. a share cash, and the million dollars therefore has been paid into the treasury of New-Kanawha Power Company. At the hearing on October 29, 1928 before the Commission, Mr.

L.H. Davis testified in answer to a question by Mr. Livezey as follows: "there has already been placed in the treasury of the New-Kanawha Power Company the sum of one million dollars through the purchase of 40,000 shares of stock of the New-Kanawha Power Company by the Union Carbide and Carbon Corporation at the price of $25.00 per share."

At the request of Mr. Livezey made at the said hearing, the Secretary of New-Kanawha Power Company prepared a certificate under date of November 15, 1928, setting forth an excerpt from the minutes of the New-Kanawha Power Company stating that the offer of Union Carbide and Carbon Corporation to subscribe for 39,995 shares of stock of New-Kanawha Power Company at the price of $25 per share payable in cash was accepted. I sent the said certificate to Mr. L.H. Davis, and he is of the impression that the certificate was given to you for filing with the Commission. The other five shares to make up the 40,000 shares are held by the Director of New-Kanawha Power Company.

There has been no further permanent financing by New-Kanawha Power Company. Amounts necessary for the construction of the hydro-electric project have been advanced on open account to New-Kanawha Power Company by one of the other subsidiaries of Union Carbide and Carbon Corporation. We shall, of course, be pleased to present to the Commission, at a formal hearing, if it shall so desire, the information given at the said hearing on October 29, 1928, and such further facts with respect to the financing of New-Kanawha Power Company as the Commission may wish to have presented. As you know, the books of the company are at all times open to the Commission and it has been the practice of the Commission to send an auditor each year to New York City to examine the books of the company and to take therefrom all information that he wishes. We assume from Mr. Coffman's letter that perhaps the Commission would prefer to have us present the financial figures by sworn testimony at a formal hearing in corroboration of the figures that the auditor of the Commission has taken from the books of the company.

2. We appreciate the fact that the acquisition of lands and other rights by New-Kanawha Power Company is subject to the approval of the Commission as to price and terms. The Commission has by an order approved the site of the hydro-electric development. Most of the lands and rights embraced in the hydro-electric site were at the time of the said order owned by Electro Metallurgical Company. However, at the time of the said order, there were a few small parcels that were owned by other parties, and it was our intention to have Electro Metallurgical Company acquire these other parcels so that it could convey a complete development site to New-Kanawha Power Company. It has taken some time to secure these small parcels but it is my opinion that all the lands and rights necessary for the development of the project included in the permit have now been acquired.

It is, and always has been, our intention to submit to the Commission for its approval, any proposed conveyance of the permit site by Electro Metallurgical Company to New-Kanawha Power Company. It is not an easy matter to fix a price for such land and rights. There are several bases upon which the determination of the value of such land and rights can be made. The conveyance would be from one subsidiary of Union Carbide and Carbon Corporation to another of its subsidiaries. There would be no trading at arms length. Condemnation would be a costly and useless gesture. There is no inclination on our part to have the lands and rights transferred to New Kanawha Power Company at an excessively high figure. We would probably suggest a low price, so that such price could, without question, be satisfactory to, and would receive the approval of, the Commission. However, any conveyance of the said land and rights involves the question of the lien of the general mortgage of Electro Metallurgical company which covers the lands included in the hydro-electric development site.

[omission]

Very truly yours,
.s. Ben. G. Smith.

LAW OFFICES[108]
CLYDE BEECHER JOHNSON
CHARLESTON, WEST VA.

August 3, 1932

Hon. I. Wade Coffman, Chairman,
Public Service Commission,
City.

Dear Sir:-

You will remember that when we discussed the New Kanawha Power Company matter and particularly the recent letter from Mr. Benjamin G. Smith in respect to this matter, it developed that the certificate of the Secretary of the New Kanawha Power Company, adopted January 17, 1927, showing subscription by Union Carbide & Carbon Corporation for 39,995 shares of the capital stock of the New Kanawha Power Company, and which had been requested at the hearing of the case in November 1928 to be filed with the Commission, did not seem to be on file, so oI requested Mr. Smith to send a new certificate so that it could be supplied and that the file in that respect would be complete. Accordingly, I am enclosing certificate as above indicated, bearing

---

108. Ibid.

date November 15, 1928, showing this subscription and the resolution accepting the same by the New Kanawha Power Company.

Will you please let this certificate be filed in the papers of the case and acknowledge receipt thereof.

I am quoting the following from Mr. Smith's letter dated August 1, 1932, respecting the recent visit of Mr. Williamson, Chief Accountant of the Commission, who called upon the accounting department of the New Kanawha Power Company, and who statement indicated that our company took the position that there was no water power act in effect in West Virginia:

"One day last week Mr. Williamson, chief accountant for the Public Service Commission, called at our Accounting Department with respect to the accounts of New Kanawha Power Company and whether or not exact costs were being kep, and similar matters. I was away from the City at the time but I am assured that our Accounting Department endeavored to give Mr. Williamson all the information that he desired. I am told that Mr. Williamson seemed to feel that detail records were being properly kept so that he would have no great difficulty in going over them at any time that he wished to make an exact audit.

"I am told that Mr. Williamson seemed to be of the impression that we were taking the position that the 1929 law having been declared unconstitutional and the 1915 law having been repealed by the 1929 law, there was no water power act in effect in West Virginia.

"We have not taken any position with respect to whether the 1915 law or any other water power act is in effect in West Virginia, nor have we taken any position that is opposed to the jurisdiction of the Public Service Commission. Probably Mr. Williamson was not correctly informed as to our attitude. However, if the Commission should be of the opinion that we are at all inclined to question the jurisdiction of the Commission, I think that they should be told at once that such is not the case.

"I am told that Mr. Williamson also commented that apparently we had violated the Commission's restrictions when at the end of 1931 we transferred the plant, which had been constructed up to that time, from New Kanawha Power Company to the Electro-Metallurgical Company; that he felt that this was a sale of the property which under the rules could not be made without the approval of the Commission."

You, I think, understand that the New Kanawha Power Company has not taken any position as to what water power act, if any, is now in effect in West Virginia, and has always freely recognized the jurisdiction of the Public Service Commission over its operations. I am sending you this copy to comply with the suggestion of Mr. Smith that your Commission should be assured upon this point.

Incidentally, I have not heard from you that the date for further hearing in this case as to the financing, etc., has been yet fixed. Will you please let me have notice of this as soon as it can be arranged by the Commission, as I would be glad to advise Mr. Smith of this date in plenty of time so that he can be prepared to bring with him all of the records and information that the Commission will desire upon the further hearing of this case.

With kind personal regards, I remain

>Yours very truly,
>Clyde B. Johnson

FEDERAL POWER COMMISSION[109]
WASHINGTON

Frank R. McNinch, Chairman
Basil Manly, Vice Chairman
Herbert J. Drane
Claude L. Draper

December 11, 1933

-DI-#102-West Virginia
New-Kanawha Power Company.

New-Kanawha Power Company
Carbide and Carbon Building
30 East 42nd Street
New York, N. Y.

Gentlemen:

Reference is made to the Commission's letter under date January 13, 1933, advising you (1) that the Commission had reached the conclusion that the action of the [FPC's] Executive Secretary, in closing the inquiry upon your declaration of intention for a power project on New River at Hawks Nest, West Virginia, was unauthorized and based upon an erroneous interpretation of the statute, that said declaration of intention was held to be still pending and would be regularly passed upon by the Commission; and (2) requesting to be informed as the manner and extent to which the construction then being carrion on by you conformed to the plans originally submitted in said declaration.

Under date of January 21, 1933, you acknowledged receipt of this communication, but failed to comply with the request contained therein. On April 15, 1933, this Commission again asked to be advised as to the manner and extent to which the construction conformed to the original plans. No answer has been received to this request.

---

109. Ibid.

Now you are advised that the Commission will on December 20, 1933, proceed to make a determination upon said declaration of intention in accordance with Section 23 of the Federal Water Power Act. If you have any further information, evidence or representations to make in the matter, they should be before the Commission on or before that date, and if you so desire, the Commission will hear oral argument on the questions raided by the declaration at 10 a.m., December 20, 1933, at the Commission's offices in this City.

Very truly yours,

Frank R. McNinch
Chairman

FEDERAL POWER COMMISSION[110]

Frank R. McNinch, Chairman
Basil Manly, Vice Chairman
Herbert J. Drane, Commissioners
Claude L. Draper

DECLARATION OF INTENTION
New-Kanawha Power Company (DI-102)

The following action was taken:

WHEREAS, The New-Kanawha Power Company on May 10, 1927, filed a declaration of intention to construct a water-power project on the New River in Fayette County, West Virginia; and

WHEREAS, the project works outlined in the declaration of intention as "first alternative" comprehend the development of a head of 148 feet, and consist of a dam 40 feet high near Hawks Nest, West Virginia; and tunnel about 3 miles long with suitable head works at its entrance and a surge chamber and distributing penstocks at its outlet; a power house on the river bank at the outlet of the tunnel about opposite Gauley Junction; and appurtenant works; and the project works described as "second alternative" comprehend the utilization of the same head in a two-stage development instead of a single-stage development as described above; and

WHEREAS, upon filing of said declaration of intention the Commission caused an immediate investigation of such proposed construction to be made, as required by the Federal Water Power Act, and

---

110. Ibid.

WHEREAS, the main stream on which this project is located is commonly called New River above the mouth of Gauley River, and Kanawha River below that point; both the New and Kanawha Rivers having been improved for navigation by the United States; and

WHEREAS, in periods of extreme low water the New River furnishes more than 95 per cent of the water which flows into the Kanawha River; and

WHEREAS, the Kanawha River, in the 90-mile section from its mouth to a point near Kanawha Falls, has been canalized by the United States, by the construction of locks and dams, for 6-feet draft navigation; and the United States is now engaged in building navigation structures for the purpose of increasing the navigable depth of this section of the river to 9 feet; and

WHEREAS, the site of the proposed power house is only 8 miles upstream from the head of the pool created by navigation dam No. 2, and only 4 1/2 miles above the head of the pool to be created by the London navigation dam now being constructed by the United States; and

WHEREAS, the investigation which the Commission has caused to be made as required by the act shows that the proposed construction would affect navigation on the Kanawha River; and

WHEREAS, a large amount of interstate commerce originates along, and is carried on, the 90-mile canalized section of the Kanawha River hereinbefore described.

Now, therefore, having considered the declaration of intention, together with reports, correspondence, and other information pertaining thereto, the Commission finds:

<u>That the interests of interstate commerce would be affected by such proposed construction.</u>

I, RALPH R. RANDELL, Acting Secretary of the Federal Commission, hereby certify that the foregoing is a true and correct of a portion of the minutes of a meeting of the Federal Power Commission in the City of Washington, District of Columbia, on the 26th day of January, 1934.

In testimony whereof, I have hereunto set my hand and caused the seal of the Federal Power Commission to be affixed in the City of Washington, District of Columbia, this 5th day of February 1934.

Ralph R. Randell
Acting Secretary

WESTERN UNION[111]

NEWCOMB CARLTON, PRESIDENT J.C. WILLEVER, FIRST VICE-PRESIDENT

Received at 803 Kanawha Street, Charleston, W. Va. 1934 DEC 20 PM 10 21

JB333 89 GOVT NL 1 EXTRA =BM WASHINGTON DC 20

CHAIRMAN JOHN J D PRESTON=
PUBLIC SERVICE COMMISSION OF WESTVIRGINIA
CHARLESTON WVIR=

IN CONNECTION WITH HAWKS NEST LITIGATION THIS COMMISSION URGENTLY NEEDS CERTIFIED COPIES OF MAPS PLANS SPECIFICATIONS AND GUARANTEES REQUESTED IN MY LETTER NOVEMBER THIRTIETH STOP

IN ADDITION TO AFORESAID MATERIAL MR LOVE OF DISTRICT ATTORNEYS OFFICE WILL CALL AT YOUR OFFICE TO OBTAIN CERTIFIED COPIES OF ALL APPLICATIONS AND AMENDMENTS THERETO TOGETHER WITH MAPS PLANS AND SPECIFICATIONS FILED BY NEW KANAWHA POWER COMPANY WITH YOUR COMMISSION INCLUDING FINAL AMENDED APPLICATIOON FOR LERMIT [sic] TO CONSTRUCT HAWKS NEST PROJECT STOP HOPE ALL OF THESE CERTIFIED PAPERS CAN BE FURNISHED PROMPTLY=

FRANK R. MCNINCH CHAIRMAN FEDERAL POWER COMMISSION

TELEGRAM[112]
OFFICIAL BUSINESS—GOVERNMENT RATES

December 26, 1934

HONORABLE JOHN J. D. PRESTON, CHAIRMAN,
PUBLIC SERVICE COMMISSION OF WEST VIRGINIA,
CHARLESTON, WEST VIRGINIA.

RETEL DECEMBER TWENTYFIRST [SIC] STOP UNDERSTAND YOUR COMMISSION WILL CALL UPON ITS PERMITTEE FOR

---

111. Ibid.
112. Ibid.

MANUFACTURERS GUARANTEES AND OTHER THINGS REQUESTED BY THIS COMMISSION AND NOT IMMEDIATELY AVAILABLE IN YOUR OFFICE STOP IS THIS UNDERSTANDING CORRECT STOP THIS COMMISSION URGENTLY NEEDS MAPS PLANS SPECIFICATIONS AND GUARANTEES REQUESTED IN PREVIOUS COMMUNICATIONS STOP YOUR COOPERATION IN OBTAINING THEM GREATLY APPRECIATED.

FRANK R. MCNINCH,
CHAIRMAN, FEDERAL POWER COMMISSION

January 15, 1936[113]

Dear Dr. Harless:

I see from the press that steps are being taken to have Congress investigate the Gauley Bridge disaster. I enclose clippings from the Columbus Citizen as of January 10th, 14th and 15th.

Certainly you are in a key position to state the chief medical facts and my best wishes go with you. I will no doubt be called later myself.

Incidentally, the American Public Health Association has addressed me a letter to ask you if you will not renew your membership in the Association. I will, of course, leave this to you, but, other things being equal, I am sure we want you to continue in the Section on Industrial Hygiene if you possible can.

Yours sincerely,
Emery R. Hayhurst

January 24, 1936[114]

Dear Dr. Harless,

I have yours of the 21st with enclosure of carbon of a letter to Hon. Wm. H. Conners [sic], and am very glad to hear from you. I also agree with you that the press has added considerable color to the news from Gauley Bridge, unless a lot of new evidence has been accumulating since my time there.

I assure you that the thought never occurred to me that you were in any way responsible for the new publicity and agitation. When newspaper men have called

---

113. From the private papers of Dr. Harless.
114. From the private papers of Dr. Harless.

up me, I have referred them to the court records at Fayetteville and to you, and that I could not add anything more. It seems to me that we have already proceeded in a dignified and scientific manner to present the situation as was done at the Indianapolis Meeting of the APHA in October, 1933, and the subsequent papers published in the American Journal of Public Health for December, 1933. Personally, I wondered why the Public Health Service and the Bureau of Mines held aloof and published nothing about the tunnel affair—at least nothing which has come to my attention.

Evidently a new weekly publication, called The People's Press, is the agency chiefly behind the present agitation, and its publisher, Mr. Frank L. Palmer, decided to use it as one of the major propositions to further his publication. One can readily see that the Gauley Bridge tragedy lends plenty opportunity for emotional outbursts. I have no idea of the make-up of the personnel of. I have seen only one representative from the paper, a printer, Mr. T.E. Silvey, who works in Columbus and is desirous of becoming a newspaper correspondent and does this work outside of his regular job. He first introduced himself to me on December 11, 1935, and started a chain of correspondence going which has involved no little part of my time and thought. He is an entire stranger to me, but appears to be a good stolid American, about 35 years old. As above stated, I have consistently referred him, as well as others who became interested after the press reports appeared, to Fayetteville and Gauley Bridge. I have no method of ferreting out Communism, except possibly to note the racial-complexion from the character of the family name. In order to size up the situation better, I paid $1.00 for a year's subscription to The People's Press. I must say, dismissing the Gauley Bridge propaganda, it is wielding a mighty club against the patent medicine interests and irregularities in hospital organizations so that I have sometimes thought it was almost a mouth-piece for the American Medical Association. Mr. Silvey says it is simply a paper to champion the rights of the "under-dog" and aim at the level of the tradesman and laborer.

Of course, I have no personal knowledge of the conditions in the tunnel, which has seemed to disappoint most of those who have interviewed me. I have told them all that I appeared as an expert only and made no local investigations.

Since they have managed to get the Congressional House Committee on Labor interested, I have received a request from Cong. Connery to appear, but I immediately inquired whether the Committee had any provisions for bearing traveling expenses and per diem fees, and, although a week has passed, I have received no response to my inquiry. Instead Mr. Silvey called upon me, and Mr. Palmer, the Editor, wrote me a letter, both hoping that I would appear. Likewise, Dr. John B. Andrews (Ph.D.), whom I have known for twenty-five years as Secretary of the American Association for Labor Legislation, has written me two letters hoping that I would appear. I have asked him the same as I asked Cong. Connery.

As with the original cases, I am perfectly willing to offer any appropriate body the benefit of my consulting services, so far as I can, and if they request same. Thus far, the only appropriate bodies with whom I would deal and who have so far appeared in the matter would be the duly constituted House Committee on Labor and the American Association for Labor Legislation, if Dr. Andrews would agree to head up any investigation which the Association might care to foster.

I would state that the American Association for Labor Legislation is not a labor organization, but an intellectual group which busies itself especially in drafting labor legislation, in an endeavor to have standard laws adopted in the various States. It drafted the original Lead Law and Occupational Disease Reporting Law under which we operate in Ohio. These laws have shown themselves to be very effective, so far as they go, in bettering labor conditions and preventing occupational diseases in Ohio. Hence, I have due respect for this Association and Dr. Andrews as its Executive Secretary. Its officers and general advisory council are composed of leading educators, sociologists, lawyers, government labor officials, and big business men in the country.

I think you proceeded quite adequately in your letter of January 19th to Cong. Connery. You, of course, have factual data immediately at hand and I would think that that is all the Labor Committee is interested in. In contrast to what the press alleges has been transpiring since I was down there in 1933, I am happy to note that you say that there are only a few cases of silicosis now in the community and these are only moderately affected. I had begun to fear that my announced belief that silicosis reached its acme within two years following last exposure and then become quiescent, in a large majority of cases, was inaccurate, or at least would not apply to acute silicosis. Your statement certainly tends to reassure me on this point.

I am enclosing a few newspaper clippings from local sources, some of which pay you a splendid compliment, and I am sure you are deserving of it.

With very best wishes, I am

   Yours sincerely,
   Emery Hayhurst

[1956]

PUBLIC SERVICE COMMISSION[115]
OF WEST VIRGINIA
CHARLESTON

At a session of the PUBLIC SERVICE COMMISSION OF WEST VIRGINIA, at the Capitol in the City of Charleston on the 13th day of December, 1956.

---

115. Public Service Commission. 75-PSC-21. Case 1863-1864. MicroFile #436B.

CASE NO. 1863

ELECTRO METALLURGICAL COMPANY,
SUCCESSOR AND ASSIGNEE OF
NEW-KANAWHA POWER COMPANY,
Application for authority to construct dam,
Hydro-electric plant and tunnel on New River
In Falls District and Kanawha District, Fayette
County, West Virginia. (Gauley Junction Dam).

Upon examination of the record in this case it is noted that the last order was entered herein approximately twenty-eight years ago. The Commission order of December 8, 1928, directed that this case be "...retained on the docket and continued for such future proceedings as are proper in the premises."

There being no future proceeding had in this case for some twenty-eight years the Commission is of opinion and finds that no useful purpose would be served by retaining this case on the docket as a continued case; that this case should be discontinued and the case removed form the docket; and it is so ordered.

[1956]

PUBLIC SERVICE COMMISSION[116]
OF WEST VIRGINIA
CHARLESTON

At a session of the PUBLIC SERVICE COMMISSION OF WEST VIRGINIA, at the Capitol in the City of Charleston on the 13th day of December, 1956.

CASE NO. 1864

ELECTRO METALLURGICAL COMPANY, successor and assignee of New-Kanawha Power Company, Application for authority to construct dam, hydro-electric plant and tunnel on New River In Fayette County, West Virginia. (Hawks Nest Dam)

This proceeding came on to be heard this 13ty day of December, 1956, upon the orders heretofore entered and proceedings had.

And examination of the entire record in this proceeding reveals that this case was "...retained on the docket and continued for such future proceedings as are proper in the premises," by order entered herein on October 2, 1929. On December 5, 1929,

---

116. Ibid.

March 18, 1930; March 20, 1930; April 5, 1930; May 27; 1930, May 11, 1931; April 24, 1933; July 31, 1933; August 5, 1933; October 2, 1934; and December 5, 1934, additional orders were entered herein. Since the order of December 5, 1934, the only activity in this proceeding has been in the form of correspondence, the latest being dated May 23, 1940.

Upon consideration whereof the Commission is of opinion that no useful purpose is served by letting this case stand continued and remain on the docket.

It is, therefore, ordered that this proceeding be, and the same hereby is, discontinued and removed from the docket.

It is further ordered that the aforesaid discontinuance and removal from the docket is without prejudice to any party of interest and is subject to the specific condition that the Commission may upon request or upon its own motion reopen and reconsider to such extent it deems necessary the matters involved herein.

A copy.

COMPUTATION OF ANNUAL CHARGE[117]
FOR THE HAWKS NEST HYDRO-ELECTRIC PROJECT
OF NEW RIVER IN WEST VIRGINIA

Chapter 115 of the Acts of 1933, Regular Session, of the Legislature of West Virginia, entitled "An Act authorizing the sale and transfer to Electro Metallurgical Company, its successors and assigns, by New-Kanawha Power Company," of the Hawks Nest Hydro-Electric Project, sometimes called the Hawks Nest-Gauley Junction Project, contains, in Section 2 thereof, the following provision:

"(b) In lieu of the annual royalty provided for by section twenty-three, chapter seventeen, acts of the legislature, one thousand nine hundred fifteen, there shall be payable during the term of the said permits, and subsequent operation thereunder, as herein provided for, an annual charge computed at the rate of ten dollars for each one hundred horsepower of water wheel capacity of the said project, which capacity shall be calculated as the product of: (1) the average stream flow in cubic feet per second at the intake; (2) the average static head in feet; (3) the factor eight one-hundredths. Such charge shall become payable beginning with the calendar year following the starting of operation of the said project and shall be made to the state tax commissioner annually thereafter prior to March one...."

The project referred to in said Act is now owned by Electro Metallurgical Company. The starting of operation of said project was on July 9, 1935, and the annual charge

---

117. Ibid.

became payable beginning with the calendar year 1937.

Such annual charge has been computed in accordance with the provisions of said Act and the basis of such computation is as follows:

There has been no gauging of the stream flow at the intake of the Hawks Nest Hydro-Electric Project and, therefore, for the purpose of calculating the power upon which the annual charge is to be based, it is necessary to compute the average stream flow on the basis of that at the nearest rating stations upstream from the intake where it has been observed for a period of years. Accordingly, the discharge figures for the present purpose are derived from those published by the United States Geological Survey for the rating station at Fayette, about 5.63 miles upstream from the intake, in its Water Supply Paper 536, entitled "Surface Water Supply of the New-Kanawha River Basin," and from those published by the Survey for the rating station at Caperton, about 9.4 miles upstream from the intake, in its Water Supply Papers 683, 698, 713, 728, 743 and 758, entitled "Surface Water Supply of the United States, Ohio River Basin," and from other as yet unpublished data on file in the office of the Geological Survey.

The published records available at the Fayette Station cover periods from July 26, 1895 to May 31, 1901; from August 11, 1902 to December 31, 1904; and from July 20, 1908 to September 30, 1916, all dates inclusive. The records at the Caperton Station began on November 20, 1928 and are available in published form to September 30, 1934. The gaugings after that date have not been published but are on record to September 30, 1936 in the office of the Geological Survey.

The records that we have used are the whole years beginning on the 1st of October and ending on the 30th of September, as reported in the water supply papers, and on file in the Geological Survey office; in two instances fractional years have been completed with estimates for the months for which records are not available.

The watershed area above the Fayette rating station is stated by the Geological Survey in Water Supply Paper 536 to be 6800 square miles. In the water supply papers giving the Caperton records the watershed area above the Caperton rating station is given as 6830 square miles. The figures are obviously inconsistent with each other because Fayette is downstream from Caperton and the watershed area between the two stations amounts to 26 square miles. The explanation is that the Caperton reports, being of a later date, are based on more accurate determinations of area. For the present calculations the area tributary to the Caperton rating station is taken, as given by the Geological Survey as 6830 square miles; the area tributary to the Fayette Station is taken, with the indicated correction, as 6856 square miles. The watershed area drained by the river above the intake of this development is found by measurement on the topographic maps of the Geological Survey to be larger than the corresponding area above Fayette by 57 square miles, the total area drained, then, being 6913 square miles at the intake at the Hawks Nest dam.

On the assumption that the average yield of water is the same from every square mile of the whole watershed above the intake, the average stream flow at the intake would be calculated by dividing the average stream flow at Fayette by 6856 and multiplying the quotient by 6913, or by dividing the average stream flow at Caperton by 6830 and multiplying the quotient by 6913. In this manner the average stream flow at the intake is calculated, as shown in the following table, to be 9,266 cubic feet per second:

STREAM FLOW CUBIC FEET PER SECOND

Years (October 1–Sept. 30)	Water Supply Paper	At Fayette	At Intake
1895-1896	536	8160	8228
1896-1897	"	10,800	10,890
1897-1898	"	7,560	7,623
1898-1899	"	11,700	11,797
1899-1900	"	6,400	6,453
1900-1901 (a)	"	16,314	16,449
1902-1903	"	10,300	10,385
1903-1904	"	5,480	5,525
1908-1909	"	10,900	10,990
1909-1910	"	7,750	7,814
1910-1911	"	8,090	8,157
1911-1912	"	10,700	10,789
1912-1913	"	8,610	8,681
1913-1914	"	8,180	8,248
1914-1915	"	9,260	9,337
1915-1916	"	11,500	11,595

At Caperton

Years	Water Supply Paper	At Fayette	At Intake
1928-1929 (b)	536 and 683	11,095	11,230
1929-1930	698	7,280	7,369
1930-1931	713	5,550	5,618
1931-1932	728	7,530	7,622
1932-1933	743	9,300	9,413
1933-1934	758	6,027	6,101
1934-1935 (c)	not published	13,030	13,188
1935-1936 (d)	not published	11,095	11,230
Average – 24 years			9,266

(a) No record June 1 to September 30; August supplied by the average of August 1895 and 1908; September supplied by the average of September 1895, 1902 and 1908; June and July estimated.

(b) No record October 1 to November 30; these months supplied form October and November 1904 observations at Fayette.

(c) Not published; obtained from office of Geological Survey

(d) Not published; computed from data obtained from office of Geological Survey

The static head is the difference between the elevation of the water surface at the intake above the dam and the elevation of the water surface at the tailrace where the water returns to the river from the power house, and the average static head is the difference between the respective averages of these water surface elevations.

The elevation at the intake of the pool above Hawks Nest dam will be 815.00 whenever the water supply is not greater than the amount that can be used in the turbines. When there is an excess of water it will be allowed to flow over the tops of the crest gates until the pool level reaches elevation 820.00 When there is a greater excess of water than can flow over the gates with a maximum pool elevation of 820.00 the gates will be opened to the extent necessary to keep the pool elevation from going above 820.00

In calculating the elevation of the river surface in the power house tailrace it is assumed that tailrace dredging to deepen and enlarge the waterway will have been done under and upstream from the railway bridge at Gauley Junction. When this shall have been done the lowest theoretical elevation of the tailrace surface would be the same as the elevation of the pool at Glen Ferris, namely, 651.00, and this could be only when the river discharge is zero, a condition that cannot occur. The tailrace elevation rises as the flow of the river increases.

The average elevation of the water surface in the tailrace in front of the power house, after the said tailrace dredging, is calculated to be 655.03.

The average static head is the difference between 815.91 and 655.03 or 160.88 feet.

The water wheel capacity of the project, computed on the basis of the average stream flow records for the twenty-four years above tabulated, these being the only pertinent, reliable records available, is, therefore, the product of (1) 9,266 second feet; (2) 160.88 feet; and (3) the factor, 0.08, and such product amounts to 119,257 horsepower.

The annual charge is $11,925.70

    Respectfully submitted,
    ELECTRO METALLURGICAL COMPANY

    Owen M. Jones
    Engineer in Charge

Glen Ferris, W. Va.
February 25, 1937

C: GLOSSARY OF TERMS

Adit Opening driven horizontally into the side of a mountain or hill for excavation or extraction of minerals.

Benches Created as result of the stair step technique utilized in drilling and excavating the tunnel.

Heading The part of a tunnel excavation where drilling occurs.

Manifold Nautilus-shaped pipe located at powerhouse-end of tunnel, with lateral outlets connecting to each of the four penstocks.

Penstocks 14' pipelines that carried the tunnel's water to individual turbines.

D: GLOSSARY OF NAMES

Coffman, I. Wade — Chairman, Public Service Commission of WV

Conley, C.A. — Sheriff of Fayette County

Conley, Joseph G. — Former Governor of WV; Secretary for the WV Public Service Commission

Connery, William P. — Chairman, U. S. Congress Committee on Labor, 1936

Davis, L.H. — Vice-president, New-Kanawha Power Company

Eary, J.W. — Fayette County judge during the silicosis trials in Fayetteville

Faulconer, P.H. — President, Rinehart & Dennis Company

Haggerson, F.S. — President of New-Kanawha Power Company when it was dissolved in 1935

Harless, L.R. — Fayette County physician who first diagnosed silicosis

Hatfield, Col. J.T. — Chairman, Hatfield-Cambell Creek Coal Co.

Hayhurst, Emery — Physician and consultant with the Ohio State Department of Health in occupational disease

Holt, Homer "Rocky" Adams — Attorney General WV. General counsel, director and vice-president of Union Carbide Corporation

Holt, Rush Dew — Elected as WV's representative to the U. S. Senate, in 1934, at age 29, and participated in the Congressional hearing on silicosis

Johnson, Clyde Beecher — Charleston-based attorney for New-Kanawha Power Company

Jones, O.M. — Chief engineer of New-Kanawha Power Company

Lambie, Robert M. — Chief, State Mines Department

Livesey, F.M. — Counsel, WV Public Service Commission

McNinch, Frank R.	Chairman, Federal Power Commission
Perkins, E.H.	Vice-president, Rinehart & Dennis Company
Ricks, Jesse	President, Union Carbide Corporation
Settle, J.E.	Engineering consultant for the WV Public Service Commission
Smith, Benjamin G.	NYC-based counsel for the New-Kanawha Power Company
Williamson, E.V.	Statistician, WV Public Service Commission
Plaintiffs' Attorneys	W.E. Teubert L. Burke O'Neal Love & Love F.N. Bacon A.A. Lilly Townsend, Bock & Moore J.K. Edmundson C.T. Dyer A.L. Russell A.M. Mason C.R. Summerfield
Defendant's Attorneys	C.W. Dillon W.L. Lee George Couch: Brown, Jackson & Knight Dillon, Mahan & White
Jurors, Raymond Johnson Trial	R.T. Saunders, Sewell A.F. Bell, Victor Robert Dice, Nuttallburg G.H. Berry, Backus B.C. Goins, Backus James Sowder, Ansted W.H. Reid, Scarbro S.F. Sanford, Bellwood Emery Kincaid, Kincaid T.A. Dodson, Meadow Bridge Mack Gill, Ansted

E: DIMENSIONS & SPECIFICATIONS, HAWKS NEST DAM & TUNNEL[118]

HAWKS NEST DAM

Length: 948 feet
Elevation at Gate Crest: 820 feet
Original elevation of New River at low water: 746 feet
Altered elevation of New River at low water: 760 feet
Discharge capacity: 350,000 cfs

HAWKS NEST TUNNEL

Length: 16,240 feet
Shaft 1 length: 4100 feet
Shaft 2 length: 3150 feet
Shaft 3 length: 3700 feet
Shaft 4 length: 5300 feet
Penstock diameter: 14 feet
Total grade: 163 feet

---

118. Cherniack, pp. 171-172.

INDEX

A. Guthrie and Company 152
Abernathy, Henry 150Agricola 62
Agsten, H.G. (and Company) 129
Air Hygiene Foundation 181
Aldridge, R.S. 146
Allen, Philippa [Bernard] 1, 2, 6, 9, 11-14, 16, 17, 21-27, 31, 45, 91, 92
Alloy, WV 19, 81, 92, 130, 135, 143, 149, 168, 170
American Smelting & Refining 182
Amick, R.E. 146
Anders, Clev 34
Anders, Oscar 34
Andrews, Sidney 17
Andrews, Thelma 17, 18
Ansted, WV 146, 147, 153, 154
Ardery, Major A.D. 113, 114
Armentrout, H. 119, 120
Armour, Wilbur 158
Armour, Williams 159
Armstrong, Lannie 146
Arthur, J.W. 154
Ashland, KY 151, 153
Atlanta, GA 138, 155
Austen, John 157

Babcock & Wilcox 161
Baber, W.F. 146
Backus, WV 146, 147, 153
Bacon, F.N. 147, 149, 152, 153, 156, 158
Ballard, S.H. 158
Ballard, W.R. 144
Barium Reduction Company 118
Beale, Warren 146
Beckett, W.M. 154
Beckley, WV 81, 144
Beckwith, WV 88
Bell, A.F. 146, 147, 153
Bell, John 159
Bellwood, WV 146, 147, 153
Berry, G.H 147, 153
Bhopal, India 183
Bishoff, C.B. 152
Blackwell, Dr. Lisle xii
Blaine Island 115, 122
Blake, Tim 146
Blankenship, Ballard 145

Blue Ash Tunnel 116
Bock, Attorney Edward S. 15, 19, 39, 157, 159, 160
Boeing 182
Boggs, William 145
Boley, J.E. 147
Bolton, John 137
Boncar, WV xi, 116-118, 121, 122, 124, 127, 129, 130, 135
Boomer, WV 146
Bostic, Edgar 157
Bostic, Lewis 154
Boxley, W.C. 19
Boyd, Richard 157
Bradford, George 158
Branham, Sam 150
Brazie, Frank H. 159
Britt, A.L. 145
Brookwood Labor College 1
Brown, B.M. 154
Brown, Ernest 159
Brown, George 159
Brown, Jackson & Knight 147, 158
Brown, John 103
Brown, Parker 138
Brown, Will 158
Brush Beryllium 182
Bryson, James 159
Bryson, Johnnie 157
Buckley, R.E. 137
Bureau of Mines (US) 73, 95, 102
Bureau of Mines (WV) 43, 63, 64, 95
Burgess, Arnold 154
Burning Springs, WV 114
Bush, Fred 154
Butler, Charles 145
Butner, Sam 4

C&O Railroad 19, 24, 82, 92, 116, 118, 147
C.C.B. Coal Company 144
C.H. Meade Coal Company 144
Campbell, C.W. 154
Campbell, Charles 145
Campbell, J.C. 154
Capone, Al 66
Carbondale, WV 154
Cavendish, Stanley 77-81

Cedar Grove, WV 1, 92
Chapman, Lonnie 159
Charles Town, WV 63
Charles, Jessie 145
Charleston Gazette 132
Charleston, WV 19, 35, 38, 52, 57, 63, 88, 92, 111, 113, 116, 122, 129, 136, 146-148, 153, 156-160, 165-168, 174
Charlottesville, VA xii, 4, 7, 11, 65, 68, 73, 101, 123, 124, 126, 137, 146, 159, 168, 176, 180
Chattfield, Fred 154
Chattfield, Major 157
Cherniack, Dr. Martin xii
Chimney Corner, WV ix
Chisholm, L.T. 152
Civilian Conservation Corps vii
Clark, Eddie 26
Clarksburg, WV 129
Clemons, Dewey 157
Cleveland, OH 160
Clifftop, WV 154
Coal Institute 64, 67
Coal Valley Hospital 11, 17-19, 126, 161, 162, 174
Coffman, I. Wade 125, 136
Cole, A.A. 157
Cole, Earl 157
Cole, George 157
Cole, Howard 157
Cole, John 157
Cole, Wilsie 145
Coleman, Dave 158
Coleman, Horace 159
Collins, Garten 159
Colmore, James 145
Columbus, OH 57, 119, 148, 156, 178
Conley, C.A. 9, 11, 132
Conley, Gov. W.G. 129, 136, 137
Conley, James C. 130
Conley, Sheriff 86, 87, 94
Connery, William H. 56, 69, 73, 75
Consolidated Coal 182
Corliss, WV 146
Cornwall, England 62
Cotton Hill Mountain vii, ix, 88, 117, 134
Couch, George 147, 153, 158
Counts, Chester C. 118
Cox, Arthur 145
Crawford, Josh 159
Crickmer, WV 146
Crickmer, WV 154
Crumpton, Gilbert 154

Danese, WV 146
Davis, G.H. 136, 137

Davis, Leonard H. 1, 113, 114, 118, 121, 124
Davis, Rev. E. Davis 144
Davis, Willie 158
Deepwater, WV 113-115
Deets (Deitz), Mr. 37, 38
Deitz, Dennis 77-79
Dennis, Hollis 136
Detch, Mr. 52
Dice, Robert 146, 147, 153
Dickinson, Frank 34
Diehl, W.L. 154
Dietz, J.R. 146
Dillon, C.W. 146, 147, 153
Dillon, H.E., Jr. 132
Dillon, Mahan & White 154, 158
Divide, WV 146
Dixon, George 144
Dodd Undertaking 138, 154
Dodd, Dr. 146
Donne, John 183
Doom, B.G. 80
Dotson, T.A. 146, 147, 153
Dow Chemical 182, 183
Dragan, Jon 83, 84
Drain, Jno. 138
Dunbar, D.H. 146
Dunbar, Mrs. 36
Dunn, Matthew 2, 11, 12, 16, 21-25, 29, 31, 34, 35, 40, 45-48, 50-56, 58-60, 62, 66, 67, 75, 103
Dyer, C.T. 156

Eary (Erie), Judge 64, 112, 152, 153, 155, 157-160
East Bank, WV 1, 2
Eastman, Vernon 145
Eastman, Wayne 145
Edmond, WV 154
Edmundson, J.K. 156
Edna, Otis 4
Eggleston, J.W. 152
Electro-Metallurgical Company x, 3, 19, 24, 26, 87, 92, 115, 116, 118, 122, 124, 125, 127, 163, 165, 168-170, 179
Elkem-Metals Company 113
Ellison, M.B. 146
Engineering and Mining Journal 178
Engineering News-Record 177
Environmental Protection Agency 183
Epps, Thos. 154

Faulconer, Harry 137
Faulconer, Linwood 137
Faulconer, P.H. 65, 66, 68, 69, 72, 74, 103, 124, 137, 158, 171, 176, 180

Faulkner, C.M. 152
Fayette County, WV 7, 8, 10, 11, 22, 72, 113, 138, 152, 158-160, 163, 167, 174
Fayette Tribune xii, 111, 113, 115, 116, 119, 121-125, 127, 129-132, 135, 136, 138-140, 144-146, 148, 151, 152, 154, 155, 157, 160, 163-167
Fayette, WV 154
Fayetteville, WV 19, 26, 64, 65, 67, 85, 88, 94, 111, 146, 148, 153, 154, 156, 160
Feazell, W.R. 146
Federal Energy Regulatory Commission 113
Federal Power Commission xiii, 48, 112-114, 163
Finch, John W. 61-63
Finke, William 58, 59
Fitch, Mr. 34
Ford Motor Company 182
Foster, Ed. 158

Gamoca, WV 6, 17, 27, 31, 147-149, 153
Garvey, William 144
Gauley Bridge, WV xi, 1, 2, 3, 8, 9, 12, 15, 17-20, 23, 27, 30, 31, 39, 50, 56, 58, 73, 75, 86-88, 92, 94, 105, 108, 120-122, 125, 132, 134, 135, 137, 144, 146, 148, 152, 154, 156-158, 160, 167, 171, 172, 175, 176
Gauley Junction, WV xi, 2, 8, 73, 81, 91, 116, 119, 126, 160, 172
Gauley Mountain, WV viii, xi
Gay, Cecil 160
Gay, Cecil 161-162
Gee, Frank 148
Geither, Ben 157
General Electric 182
General Hospital 35
General Mills 182
General Motors 182
Geter, J.T. 157
Gibson, Harley (Harless) 8, 9, 11, 22
Gibson, Henry 157
Gilbert, WV 151
Gill orchestra 136
Gill, John 157
Gill, Mack 146, 147, 153
Gilmer, Charles 146
Gilmore, Charles M. 26, 150
Gladden, Dock 157
Gladden, Jessie 157
Gladwell, Frank 157
Glen Ferris Inn 126
Glen Ferris, WV 42, 87, 114, 115, 117-119, 122-125, 127, 129, 146, 162
Goins, B.C. 146, 147, 153
Goldwater, Dr. 99, 100

Gollman, James 154
Gooch, J. 137
Goodall Rubber Company 152
Goodyear 182
Graff, E.H. 144
Graney, P.C. 144
Green, Fred 158
Grey, Leo 18
Griffin, T.G. 146
Griswold, Glenn 2, 6, 12-14, 16, 17, 21, 24, 25, 27, 31, 34-36, 38, 39, 42-45, 49, 50-54, 58-63, 75, 99, 100, 103
Grose, J.T. 146
Gulf Oil 182
Gulf Smokeless Company 144
Gwinn, E.L. 146
Gwinn, J.E. 154

Hall, Edward 145
Hall, John W. 160, 162
Harless, C. Louise 86
Harless, Dr. L.R. 28, 32, 52, 53, 56, 58, 59, 87, 88, 93, 98, 146, 148, 155, 175
Harless, Vic 88
Harless, Walter 86-89
Harris, James 158
Harris, Jim 158
Hartman, Charles 159
Hatfield-Campell's Creek Coal Company 117
Haviland sisters 136
Hayes, Steve 157
Hayhurst, Dr. Emery R. 19, 35, 57, 59, 93, 97, 148, 156, 178
Haynes, D.W. 160
Haynes, W.H. 159
Head, George 159
Henderson, Fred 158
Henigan, Joe 94
Hess, Howard 154
Hess, W.V. 146
Hico, WV 146
Hill, Dr. D.L. 144
Hippocrates 62, 95
Hirth, Alfred C. 181
Hodger, C.A. 146
Holland Tunnel 94
Holliday, Earl 146
Hollygrove, WV 1
Holmes, Johnnie 157
Holt, Atty. Gen'l. Homer A. 163, 166
Holt, Senator Rush Dew 59, 60, 94, 175, 176
Hoover, Herbert 87
Hopkins, Ernest 159
Hopson, Ernest 158
Horner, Carl L. 129

Houston, George 8, 9, 11, 23
Howard, Charles 144
Howard, Edward 144
Howell, Henry 146
Hubert (Hubbard) and Bacon 38, 39, 54, 145
Hudson, Judge Arthur P. 164
Huey (Hughey), Dr. 32, 35, 57, 148
Huffman, J.J. 4
Hughes, Dr. 93
Hugo, Victor 74
Hugson, Frank 145
Humphrey, N.P. 146
Huntington, WV 115
Hurvitz, Ben 146

Industrial Health Foundation 182
Industrial Hygiene Foundation 182

Jackson, Henry 157
Janey, Esten 145
Jeffrey, Oley 27, 34
Jenkins, Mose 158
Jennings, Witt 87
Jodie, WV 146
Johns-Manville Corp. 182
Johnson, B.F. 138
Johnson, Clyde B. 117, 125, 137
Johnson, Elijah 145
Johnson, H.W. 146
Johnson, John 159
Johnson, Monroe 158
Johnson, Raymond 4, 17, 34, 39, 43, 93, 147-152, 156, 158, 159
Johnson, Robert 154
Jones, Cecil I. 6, 7, 17, 27, 34, 39, 52, 98, 139, 147, 148, 153
Jones, Charles 17, 20, 31-39
Jones, Dora 139, 145, 147, 154
Jones, Dorothy 39
Jones, Lindsey 18
Jones, Mrs. Charles 6, 9, 27-31, 93
Jones, Nancy 18
Jones, Oren 6, 7, 17, 27, 34, 39
Jones, Owen M. 6, 113, 117, 118, 121, 123-125, 127, 129, 137, 140, 149, 150, 158, 159, 162, 179
Jones, Shirley 6, 7, 16, 17, 27, 34, 35, 37, 38
Jones, U.S. 154
Joplin, MO 182
Jordan, Jim 157
Jumper, Cornelius 157

Kanawha Falls 115, 116, 120, 127, 146
Katonah, NY 1
Kawecki Berylco Industries 182

Keyser, Dr. 162
Kidd, Charles 159
Kies (purchasing agent) 11
Kilsyth, WV 154
Kincaid, Emory 146, 147, 153
Kincaid, WV 146, 147, 153
King Tut 62
King, Laird 4
Kingston, WV 146
Kirk, Roy 145
Kistler, John 150
Knight, E.W. 146
Knoxville, TN 138
Koppers Coal Company 9, 32
Krebs, Charles E. 118, 124
Kube, A.L. 160, 162

Lambert, Dr. 148
Lambertson, W.P. 2, 23, 27-29, 32, 36-40, 75, 103
Lambie, Robert M. 5, 6, 67, 132, 135, 147, 149-151
Lancaster, Geo. T. 125
Landisburg, WV 154
Lansing, WV 146
Lawton, WV 146
Layne's Garage 36, 86
Lee, Judge W.L. 146, 147, 153, 158
Lee, Williard 158
Leek, Mrs. 93
Legg, Chester 154
Liberty Tubes 94
Lilly, A.A. 69, 147, 153, 158
Lilly, Robert 140, 144
Lochniss 62
Long, R.L. 154
Lookout, WV 146
Love & Love, 138
Love, Frank 147
Lucas, WV 154
Lyes, Ernest 11, 138
Lynch, E.R. 144
Lynch, Frank 34
Lynch, Robert 158

Mabe, J.C. 144
Mabscott, WV 144
Madison, WV 64, 67
Mahoney, J.W. 146
Maltz, Albert 104
Maplewood, WV 146
Marby, Ivory 158
Marcantonio, Vito 2, 9, 25-27, 30-34, 37-39, 41, 43-46, 48, 53-55, 58, 62, 63, 65-68, 75, 99, 103, 167-169, 175-180

Marshall, G.F. 154
Martin, J.L. 146
Maryland New River Company 144
Mason, A.M. 154
Mason, J.S. 144
Mason, James M. 52, 53, 63, 65-68, 112
Massey, Robert 146
Maxwell, Mr. 162
Maynor, R.L. 154
McAttee, Howard 15
McCloud, Deputy Sheriff "Cap" 8, 9, 49, 51, 84
McCoy, Fred 159
McCulloh, Hansy 109
McKell C&C Company 144
McMorrow, Charles 144
Meadow Bridge, WV 146, 147, 153, 154
Meadows, Clarence 157
Meadows, J.L. 154
Means, Charles 139, 140
Means, Mr. & Mrs James 140
Mellon Institute 181
Metheney, B.H. 81-86
Midvale Colliery Company 35-38
Miller, Judge 154
Miller, Nancy 36
Mitchell, Dr. R.C. 6, 18, 149
Mobil Oil 182
Mohr, Clarence 146
Montgomery News 157-160, 163, 167, 168, 171, 175, 177-179
Montgomery, Clev 16
Montgomery, WV 11, 28, 146, 156, 157, 160, 161, 174
Moore, Ben 15, 157
Moore, Joseph 157
More, Ed. 145
Morehead, Major 125
Mosby, William 145
Moses, Harvey 159
Mount Hope, WV 6, 144, 146
Murphy, E.C. 145
Murray, Isaac 154
Murray, J.B. 154
Musick, Rev. Wm. 125
Myles, George 146

Nallen, WV 152
Neal, District Atty. Geo. I. 163
Neelwy, W.L. 154
Nelson and Chase Gilbert Company 151
Nelson, H.W. 151
New River & Winding Gulf Mining Institute 140, 144
New York City 12, 13, 41, 58, 59, 71, 122-124, 137, 163, 165, 174

New York Department of Labor 177
New York Times 177
New-Kanawha Power Company x-xii, 1, 3, 4, 6-9, 12-14, 19, 24, 26, 43, 64, 68, 73, 92, 101, 111-125, 127, 129, 132, 135-137, 139, 140, 143-145, 149, 150, 152, 154, 158-160, 163-165, 170, 178
Niagara Falls, NY 127, 135
Nicholas County, WV 11, 148, 149
Nicholas, Fayette and Greenbrier Railroad 152
Nichols, Robert 145
Nuckols, L.T. 151
Nuttallburg, WV 146, 147, 153
Nutter, R.O. 146

Oak Hill, WV 154
Occupational Safety and Health Act 182
Ohio Department of Health 19
Oliver, T. 158
Olstead, Mr. 36
O'Neal, L. Burke 138
Ongaro, Charles 146
Orangeburg, SC 18
Orr, David 81-86
Owens-Corning Fiberglass 182

Page, WV 154
Paint Creek, WV 1
Palf, Henry 34
Palmer, Willie 158
Pancoast, Dr. Henry K. 151
Parker, L.D. 146
Patterson, Andy 85
Patterson, J.H. 160, 162
Perdue, N. 119, 120
Perkins, E.J. 18, 19, 64-66, 68, 69, 72, 74, 103, 137, 140, 142, 148, 158, 159
Perkins, Frances 177, 181, 182
Perkins, Robert 137
Peyton, Arthur 42-50, 68, 93, 99, 100
Philadelphia, PA 151, 152
Pioneer Youth Club 2
Pitckett, Jack 109
Pitts, Curlin 154
Pittsburgh, PA 127, 129
Pliny the Elder 62, 95
Plymouth Hotel 58
Powellton, WV 146
Power Commission (WV) 3, 24
PPG Industries 182
Pratt, WV 1
Prescott, Tom 109
Price Hill Colliery Company 144
Princeton, NJ 151
Proctor, C.N. 154

Public Service Commission (WV) x, xiii, 111, 112, 116-119, 121, 124, 150
Pucel, Willie 145
Pugh, Calvin 159

Ragland, T.R. 125
Raleigh, Walter 157
Randolph, Jennings 2, 9, 12, 13, 21, 23, 31, 32, 41-44, 48, 53, 56, 63, 65-67, 75, 76
Red Star, WV 146
Reid, W.H. 146, 147, 153
Rhodesville, VA 160
Richards, Russel 119, 120
Richmond, Dr. B.B. 144
Ricks, Jesse J. 13, 14
Riley, J.J. 118
Rinehart and Dennis xi, xii, 4, 6, 7, 10, 11, 13, 14, 17-19, 35, 37-41, 43-47, 54, 55, 58, 63-69, 72-75, 84, 101, 111, 112, 123, 124, 126, 131, 132, 135, 137-140, 144-159, 164-166, 168, 169, 171, 176, 180
Rinehart, Hollis 137
Robertson, W.A. 154
Robinson, Will 150, 159
Robison, George 50-56, 93, 94
Romont, WV 146
Roosevelt, Franklin Delano 181
Rucker, Dr. 40
Ruckeyser, Muriel 91
Russell, A.L. 154
Russell, Charles 150
Russell, Doc. 138
Russell, James 159
Russelville, WV 154

Sadler, Sam 158
Saler, Agnes 1
Salisbury, NC 17
Sanford, S.F. 146, 147, 153
Saunders, R.L. 146, 147, 153
Scarbro, WV 146, 147, 153
Scarbrough, Ben 138
Schadel orchestra 136
Scott, Lewis 166, 167
Sewell, WV 146, 147, 153, 154
Shay, Donald 19, 20, 43, 158, 159, 164, 165
Shepherd, Howard 132
Shepherd, Percy 158
Sheppard, Thomas C. 138
Shultz, J.D. 146
Simmons, Dr. 162
Simmons, W.P. 125
Simmons, Washington 159
Simms, Alderson 149
Simms, James 145

Skaggs, Hiram 39-42
Skaggs, Joe 146
Skelton, WV 144
Skidmore, Hubert xii
Smailes, G.T. 154
Smith, Attorney 13
Smithers, WV 154
Snyder, P.M. 144
Snyder, T.H. 144
South Charleston, WV 115, 122, 124, 135
Sowares, E.J. 1
Sowder, James 146, 147, 153
Spencer, WV 154
Springdale, WV 146
St. Petersburg, FL 120
Stallard, Dr. C.W. 162
Standard Oil of New Jersey 182
Sterling Coal Company 144
Stevenson, Aaron 138
Stinson, Fred 157
Stokes, A. 137
Straughan, E.E. 146
Street, Louis Walter 148
Stringer, Ralph 160, 162
Sullivan, D.R. 152
Sullivan, Ken 77, 79-81
Sumerville, Mr. 53
Summerfield, Atty. C.R. 154
Summersville, WV xi, 11, 12, 153
Swetman, Jake 18
Swiss, WV 81
Sykes, Paul 160, 162

Tate, Albert 145
Taylor, J. Alfred 137
Teubert, Waittman E. 120, 138, 147, 153, 156, 158
Thomas, T.J. 52
Tidwell, Horace 157
Tillman, James 157
Tincher, W.C. 146
Tinsley, Juanita 93, 96
Tissue, Wm.E. 144
Tobee, Jasper 154
Townsend, Atty. Thos. C. 157
Townsend, Bock & More 9, 15, 19, 39, 53, 68, 112, 157, 158
Tugwell, Rexford 2
Tully, L.S. 130
Turner, J.J. 146
Tyler, Alicia xii

Union Carbide and Carbon Corporation x-xii, 3, 12-14, 16, 19, 59, 73, 87, 92, 100, 111-115, 117, 121, 122, 124, 125, 127, 129, 132-135, 143, 163, 165, 166, 168, 182, 183

United Mine Workers (WV) 53
University of Pennsylvania 151
US Department of Labor 181, 182
US Steel 182

Vanetta, WV 12, 15-18, 50, 92, 139, 140, 144
Vanetter, Arley B. 138
Vawter, C.E. 154
Venson, Milledge 4
Victor, WV 146, 147, 153
Virginian Railroad 151

Wade, Dewey 138
Wall, C.C. 40
Wall, Mr. 34
Ward, Samuel 144
Ward, W.M. 144
Warren, Eugene 158
Washington, DC 56, 63, 69, 73, 75, 167, 176
Watkins, George 158
Watson, Hobart 144
Watson, James 157
Waugh, Clifford C. 137, 150
Welstone, P.J. 137
Wertz, W.W. 158
West Virginia Compensation Act 60
West Virginia News 176
West Virginia Power Company 113
Western Electric 182
Westinghouse Electric and Manufacturing 127, 129
Weston, WV 105
White Oak Coal Company 144
White, H.C. 11, 12, 180
White, John 146
White, Melvin 158
Whitney, Edward S. 117
Williams, Andrew 138
Williams, Ferry 158
Wilson, James 158
Wilson, Robert 158
Winebrenner, Ruby 37
Winona, WV 146, 154
Wise, General 96
Wood, Mr. (protester) 13
Wriston, B.J. 146
Wriston, W.L. 154
Wyatt, Lee 138
Wyco, WV 144

Yant, William 62
Yates, Sam 159
Youell, B.H. 146
Young, Albert 5

BIBLIOGRAPHY

I. Transcripts from Congressional Hearings:

U. S. Congress. House. Committee of Labor. 1936. "An Investigation Relating to Health Conditions of Workers Employed in the Construction and Maintenance of Public Utilities." Hearings on House Res 449. 74th Cong., 2d sess.

II. Books and Periodicals

Cherniack, Martin. *The Hawk's Nest Incident: America's Worst Industrial Disaster.* New Haven and London: Yale UP, 1986.

Comstock, J. 1973. "476 graves." *West Virginia Heritage*, 7:1-194. Richwood, WV: West Virginia Heritage Foundation.

Cometti, Elizabeth and Festus Summers, ed. *The Thirty-fifth State: A documentary History of West Virginia.* Morgantown: WVU Press. 1966

Crandall, William and Richard Crandall. "Revisiting the Hawks Nest Tunnel Incident: Lessons Learned from an American Tragedy." *Journal of Appalachian Studies.* Volume 8, Number 2. pg. 261-281.

Forman, Pat. "Scandal at Gauley Bridge." *Mountain Heritage.* August, 1978

Gauley Bridge Historical Society 1991. *A pictorial history of Gauley Bridge.* Gauley Bridge, WV: Gauley Bridge Historical Society.

Jennings, C. 1997. "Was Witt Jennings involved? The Hawks Nest tragedy." *Goldenseal*, Spring, 44-47.

Jennings, C. 1997. "Making a Life in the Valley: Witt Jennings of the Upper Kanawha." *Goldenseal.* Spring, 40-47.

Jordan, J. 1998. "Hawks Nest." *West Virginia Historical Society Quarterly*, 12, no. 2:1-3

National Geographic Society, 1975. *We Americans.* Washington, DC: National Geographic Historical Society.

Skidmore, H. 1941. *Hawks Nest.* New York: Doubleday, Doran and Co.

Tyler, A. 1975. "Dust to dust." *The Washington Monthly*, January, 49-58.

Schackelford, Laurel and Bill Weinberg, ed. *Our Appalachia: An Oral History. Appalachian Oral History Project.* New York: Hill and Wang. 1977.

Williams, John Alexander. *West Virginia and the Captains of Industry.* Morgantown: WVU Press, 1976.

III. Newspaper Articles:

1928

Fayette Tribune. "Hearing on Permit for Kanawha New River Power," 1/1928
_____. "Notice of Application to Construct a Dam", 6/20/28
_____. "Plans of New Kanawha Co To Use New River Power" 9/12/28

_____. "Road Building in West Va. 2241 Miles now Paved", 12/12/28
_____. "State May Take New River Dam Power After 50 Years", 12/12/28

1929
_____. "Appalachian Co. After New River Water Powers," 2/6/29
_____. "Two Engineers Drowned in Hawks Nest Rapids," 5/29/29
_____. "May Be One Tunnel," 6/26/29
_____. "One Tunnel, One Power Plant, is Revised Plan," 7/31/29
_____. "New Kanawha Company to Open Bids September 3rd on Tunnel and Dam," 8/28/29

1930
_____. "Contractors Signing Up for HN Power Project," 4/1/30
_____. "Hydro Electric Power on Kanawha," 3/30/30
_____. "Forty Engineers Working on HN Project," 1/29/30
_____. "Contractor Signing up for HN Power Project," 2/19/30
_____. "Contract Let For HN Dam and Tunnel," 3/19/30
_____. "HN Dam to be done in 2 years, cost: $4,225,737," 3/26/30
_____. "Arrest Detroit Murderer at HN Monday," 9/10/30
_____. "Award $1,000,000 contract for Big Turbine Generators," 9/24/30
_____. "First Unit Boncar Plant to be Started," 10/8/30
_____. "Glen Ferris Engineer Given Appointment on Board," 12/10/30

1931
_____. "Important Events During 1930 Reviewed from Tribune Files," 1/7/31
_____. "Control Equipment Ordered for HN Plant," 2/4/31
_____. "Charleston Firm Given Boncar Brick Contract," 2/18/31
_____. "Electro Met. Co. has Beautification Program," 4/22/31
_____. "Proposal Made to change name Boncar in petition to court," 3/11/31
_____. "Find Negro in River," 5/13/31
_____. "Army of Workmen drilling through Gauley Mountains," 6/3/31
_____. "Chief R. M. Lambie Makes Investigation at Tunnel," 5/20/1931
_____. "Ceremony Today to Mark Completion of Part Tunnel," Aug 12, 1931

1932
_____. "Ansted Negro and Wife Jailed in NY State," 4/27/32
_____. "Ventilation Study Starts a Short Course on Tuesday," 7/6/32
_____. "Another Action Against Rinehart and Dennis," 8/10/32
_____. "Negro Tunnel Worker Found Dead at Camp 2," 9/7/32
_____. "Five More Suits Filed," 9/28/32
_____. "Plaintiff in Suit Dies," 9/28/32
_____. "President Hoover Scheduled to Pass through Fayette County," 10/19/32
_____. "Text of Hoover's Address at Charleston," 10/26/32
_____. "Test Case in Damage Suits before Supreme Court," 11/23/32
_____. "First of 23 towers for Alloy plant completed," 11/30/32
_____. "Vanetta Man Another victim of silicosis, Dec. 7," 12/14/32

1933
_____. "200 Coal Men Inspect the New Kanawha Co.," 1/11/1933
_____. "Colored Gauley Bridge Man Victim of Silicosis," 2/22/1933
_____. "18 New Suits are Filed Against Construction Co," 3/1/1933

_____."Five More Suits Entered," March 8, 1933
_____."First Silicosis Case to Be Heard in Court Thursday," Volume 35, no. 15
_____."50 Additional Jurors Drawn for Circuit Court Service," 3/15/1933
_____."First Silicosis Case Gets Under Way in Court," 3/22/1933
_____."Seven More Suits Filed," 3/29/1933
_____. "Complete Second Week of Testimony in Silicosis Case," 3/29/33
_____. "Defense of Silicosis Cases Begun in Court Here,", 4/5/1933
_____. "X-Ray Expert Says Johnson Had Tuberculosis," 4/12/1933
_____. "Disagreement in Silicosis Trial: Jury is Dismissed," 4/26/1933
_____. "Tentative Date is Set for "Silicosis Trial," 5/17/1933
_____. "Silicosis Case Reviewed by Engineering Journal," 5/24/1933
_____. "Second Silicosis Case is Postponed Til June 12," 6/7/1933
_____. "More Suits Filed Against Hawks Nest Contractors," 6/7/1933
_____. "Post Mortem Performed on Body of Hawks Nest Man," 6/14/1933
_____. "Jury is Discharged in Second Silicosis Case," 6/14/1933
_____. "Negro Taken in Raleigh, Admits Killing of Teubert," 7/26/1933
_____. "Noted Physicians Getting Data on Silicosis Cases, " 7/26/1933
_____. "Silicosis Cases Settled: $130,000 is paid to Plaintiffs," 8/2/1933
_____. "Dismissal of Silicosis Cases Meets Objections," 8/23/1933

1934
Montgomery News. "More Silicosis Suits Filed," 1/4/34
_____. "Silicosis Case to Come Up Monday," 6/14/34
_____. "Fifty-One Suits Files This Week," 6/21/34
_____. "Silicosis July Fails to Reach Agreement," 7/12/34
_____. "Silicosis Suits Up for Action," 10/4/34
_____. "Kanawha Gets Silicosis Suits," 10,18/34
_____. "State Entered in Hawk's Nest License Action, Claim New River Not Navigable,"
_____. "Constitutionality of Federal Water Power Act Challenged," 12/20/1934
Fayette Tribune. "Five Die, 2 escape as power penstock bursts," 11/8/34
_____. "State to take action in HN Power Project, " 11/22/34
_____. "Holt will face fight if he appears January 3," 12/30/34

1935
Fayette Tribune. "Order in New River Suit," 1/17/35
_____."Silicosis Cases Set for April 8th at Charleston," 3/7/35
_____. "State asks dismissal of Federal HN suit," 3/21/35
_____. "Fayette Silicosis Case Postponement Ordered," 4/11/35
_____. "Supreme Court to Decide on Silicosis case soon," 4/25/35
_____. "Hawks Nest Case Today," 5/2/35
_____. "State wins HN fight; new action seen," 5/23/35
_____. "200 silicosis cases are outlawed by limitation," 5/30/35
_____. "Decision on status of Sen. Rush Holt today," 6/20/35
_____. "On seating Rush Holt," 6/27/35
Montgomery News. "Tunnel Deaths in Hawk's Nest Project Cited," 1/17/36
_____. "Tunnel Working Conditions are Given to Public, PH Faulconer," 1/24/36
_____. "Big Hawks Nest Plant to begin Operation Soon," 1/24/236
_____. "The Silicosis Racket," 1/31/36
_____. "Doctor Denies So Many Deaths from Silicosis, " 2/7/36
_____. "Silicosis New Racket, Claim," 3/20/36

IV. Literature and Music:

Allen, B. "Two thousand dying on a job." *New Masses*, 15 January, 1935.

Allen, B. "Two thousand dying on a job: 2. How the tunnel workers lived." *New Masses*, 22 January, 1935.

Maltz, Albert. "Man on a Road." *New Masses*, 1/8/1935.

Rukeyser, Muriel. *U.S. 1*. New York: Covici, Friede Publishers. 1937.

IV. Interviews:

Deitz, D. 1990. "I think we've struck a gold mine: A chemist's view of Hawks Nest." *Goldenseal*, Fall, 42-47.

Orr, D., and J. Dragan. 1981. "A dirty place to work: B.H. Metheney remembers Hawks Nest tunnel." *Goldenseal*, January-March, 34-41.

Harless, Louise. October and November, 2003. Author's collection.

V. Miscellaneous:

Air Hygiene Foundation. http://www.sourcewatch.org/index.php?title+Air_Hygiene_Foundation

"OSH Act of 1970." http://www.osha.gov/pls/oshaweb/owadisp.show_document?p_tabl

Personal letters, from the Dr. L. Harless family collection.

Personal papers, Ruby Winebrenner, Gauley Bridge, WV.

Ronald Reagan and Albert Maltz, Testimony before HUAC, 1947. U. S. Congress, House, Committee on Un-American Activities. Hearings (1947).

Silicosis Mortality, Prevention, and Control—United States, 1968-2002. The Journal of the American Medical Association. Vol. 293, No. 21, June 1, 2005.
http://jama.ama-assn.org/cgi/content/full/293/21/2585

What You Really Want Is An Autopsy.... http://historymatters.gmu.edu/d/128/

West Virginia Public Service Commission. MicroFile#436B. 75-PSC-21. Cases 1863-1864.

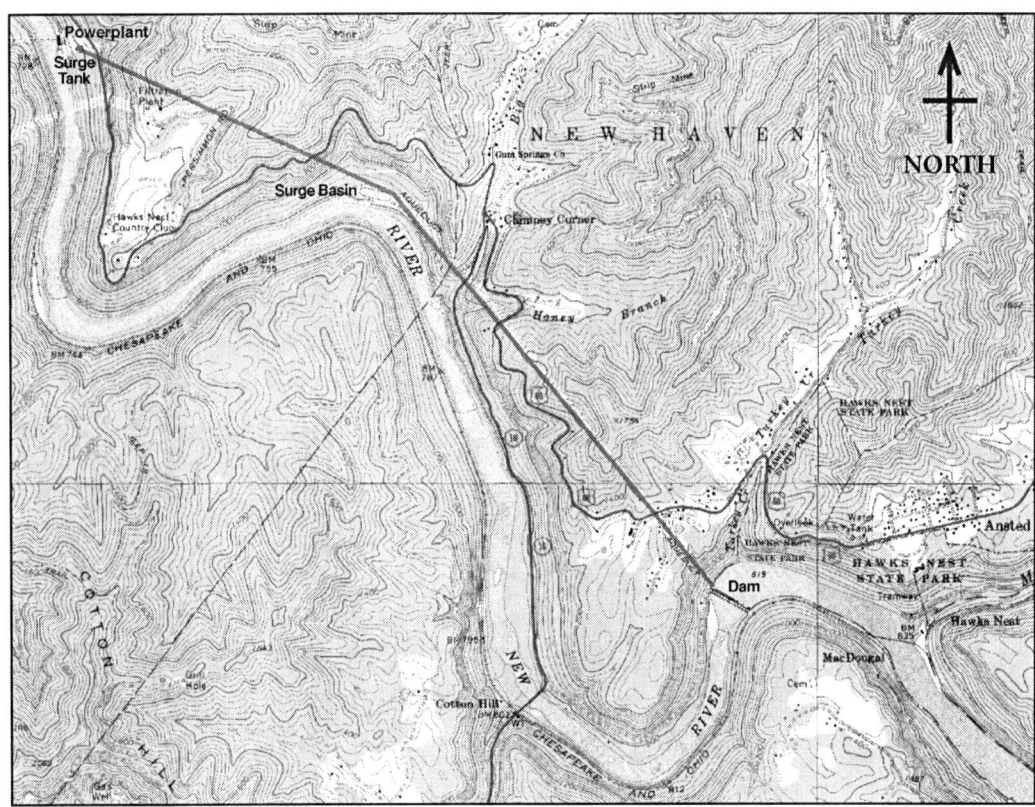

Topographical map showing the location of the tunnel relative to Hawks Nest State Park. The tunnel, marked with a dark gray line, runs a diagonal course northwest of the dam, shown in the lower right center of the map. Approximate scale: 1" = 0.68 miles. *Courtesy the U.S. Geological Survey.*

ABOUT THE AUTHOR

A native of West Virginia, Patricia Spangler and her husband live on a mountaintop farm in Fayette County.

Printed in the United States
104280LV00002B/6/P